© David Beyda Studio, NYC

Jerry Fodor is a professor of philosophy and cognitive science at Rutgers University.

D0291112

Courtesy of Fondazione Pio Manzu

Massimo Piattelli-Palmarini started his academic career as a biophysicist and molecular biologist and is now a professor of cognitive science at the University of Arizona.

ALSO BY JERRY FODOR

LOT 2: The Language of Thought Revisited

Hume Variations

The Compositionality Papers

In Critical Condition

Concepts: Where Cognitive Science Went Wrong

The Elm and the Expert: Mentalese and Its Semantics

Holism: A Shopper's Guide

A Theory of Content and Other Essays

*Pyschosemantics: The Problem of Meaning
in the Philosophy of the Mind*

The Modularity of Mind: Essays on Faculty Psychology

*Representations: Philosophical Essays
on the Foundations of Cognitive Science*

The Psychology of Language

Psychological Explanation

The Structure of Language

ALSO BY MASSIMO PIATTELLI-PALMARINI

*Of Minds and Language: A Dialogue with Noam Chomsky
in the Basque Country*

Inevitable Illusions: How Mistakes of Reason Rule Our Minds

*Language and Learning: The Debate Between
Jean Piaget and Noam Chomsky*

WHAT DARWIN GOT WRONG

WHAT DARWIN GOT WRONG

Jerry Fodor
Massimo Piattelli-Palmarini

Picador

———

Farrar, Straus and Giroux
New York

www.picadorusa.com

Picador® is a U.S. registered trademark and is used by Farrar, Straus
and Giroux under license from Pan Books Limited.

For information on Picador Reading
Group Guides, please contact Picador.
E-mail: readinggroupguides@picadorusa.com

The Library of Congress has cataloged the Farrar, Straus
and Giroux edition as follows:

Fodor, Jerry A.
    What Darwin got wrong / Jerry Fodor, Massimo Piattelli-
Palmarini.
        p. cm.
    Includes bibliographical references and index.
    ISBN 978-0-374-28879-2
    1. Natural selection. 2. Evolution (Biology)—Philosophy.
I. Piattelli-Palmarini, Massimo. II. Title.
    QH375.F63 2010
    576.8'2—dc22

                                                    2009043067

Picador ISBN 978-0-312-68066-4

Originally published in Great Britain by Profile Books Ltd.

First U.S. edition published by Farrar, Straus and Giroux

First Picador Edition: March 2011

10 9 8 7 6 5 4 3 2 1

It is perfectly safe to attribute this development to 'natural selection' so long as we realize that there is no substance to this assertion; that it amounts to no more than a belief that there is some naturalistic explanation for these phenomena.

Noam Chomsky, *Language and Mind*, 1972

I am well aware that scarcely a single point is discussed in this volume on which facts cannot be adduced, often apparently leading to conclusions directly opposite to those at which I have arrived. A fair result can be obtained only by fully stating and balancing the facts and arguments on both sides of each question.

Charles Darwin, *On the Origin of Species*, 1859

If it could be demonstrated that any complex organ existed which could not possibly have been formed by numerous, successive, slight modifications, my theory would absolutely break down.

Charles Darwin, *On the Origin of Species*, 1859

CONTENTS

DETAILED CONTENTS

ACKNOWLEDGEMENTS

For most valuable comments on a previous draft of this manuscript, we wish to thank: Bob Berwick; Cedric Boeckx; Noam Chomsky; Gabriel Dover; Randy Gallistel; Norbert Hornstein; Dick Lewontin; Terje Lohndal; Fernando Martinez; Bob Sitia; Juan Uriagereka; Donata Vercelli; Andy Wedel. The present version has greatly benefited from their suggestions and critiques. But they are not to be held responsible for errors that persist, and some of them don't agree with some of the things we say. We are grateful to Jerid Francom (University of Arizona) for his invaluable assistance in the formatting of the manuscript and in organizing the bibliography.

TERMS OF ENGAGEMENT

This is not a book about God; nor about intelligent design; nor about creationism. Neither of us is into any of those. We thought we'd best make that clear from the outset, because our main contention in what follows will be that there is something wrong – quite possibly fatally wrong – with the theory of natural selection; and we are aware that, even among those who are not quite sure what it is, allegiance to Darwinism has become a litmus for deciding who does, and who does not, hold a 'properly scientific' world view. 'You must choose between faith in God and faith in Darwin; and if you want to be a secular humanist, you'd better choose the latter'. So we're told.

We doubt that those options are exhaustive. But we do want, ever so much, to be secular humanists. In fact, we both claim to be outright, card-carrying, signed-up, dyed-in-the-wool, no-holds-barred atheists. We therefore seek thoroughly naturalistic explanations of the facts of evolution, although we expect that they will turn out to be quite complex, as scientific explanations often are. It is our assumption that evolution is a mechanical process through and through. We take that to rule out not just divine causes but final causes, *élan vital*, entelechies, the intervention of extraterrestrial aliens and so forth. This is generally in the spirit of Darwin's approach to the problem of evolution. We are glad to be – to that extent at least – on Darwin's side.

Still, this book is mostly a work of criticism; it is mostly about what we think is wrong with Darwinism. Near the end, we'll make some gestures towards where we believe a viable alternative might lie;

but they will be pretty vague. In fact, we don't know very well how evolution works. Nor did Darwin, and nor (as far as we can tell) does anybody else. 'Further research is required', as the saying goes. It may well be that centuries of further research are required.

You might reasonably wonder whether writing a critique of the classical Darwinist programme is worth the effort at this late date. Good friends in 'wet' biology tell us that none of them is '*that* kind' of Darwinist any more; no one in structural biology is a bona fide adaptationist. (Some of the reasons why they aren't will be reviewed in Part one.) We are pleased to hear of these realignments, but we doubt that they are typical of biology at large (consider, for example, ongoing research on mathematical models of optimal natural selection). They certainly are not typical of informed opinion in fields that either of us has worked in, including the philosophy of mind, natural language semantics, the theory of syntax, judgement and decision-making, pragmatics and psycholinguistics. In all of these, neo-Darwinism is taken as axiomatic; it goes literally unquestioned (see Appendix). A view that looks to contradict it, either directly or by implication, is *ipso facto* rejected, however plausible it may otherwise seem. Entire departments, journals and research centres now work on this principle. In consequence, social Darwinism thrives, as do epistemological Darwinism, psychological Darwinism, evolutionary ethics – and even, heaven help us, evolutionary aesthetics. If you seek their monuments, look in the science section of your daily paper. We have both spent effort and ink rebutting some of the most egregious of these neo-Darwinist spin-offs, but we think that what is needed is to cut the tree at its roots: to show that Darwin's theory of natural selection is fatally flawed. That's what this book is about.

In the course of it, we propose to indulge a penchant for digressions. The critique of Darwinism that we will offer raises side issues that we just can't bear not to discuss. So, we've allowed ourselves various asides we think are interesting. Our excuse is that a lot of issues that at first appear to be orthogonal to our concerns turn out, on closer consideration, not to be. We are occasionally asked whether we can really believe that we have found 'fatal flaws' in a body of theory that has been, for such a long time, at the centre of scientific consensus. We are reminded that hubris is a sin and are cautioned against it. Our reply is that, if the kinds of complaints that we will raise against Darwinism have not previously been noticed, that's partly because they have fallen between stools. It seems to us past time to rearrange the furniture. For example, we will run a line of argument that goes like this: there is at the heart of adaptationist theories of evolution, a confusion between (1) the claim that evolution is a process in which *creatures with adaptive traits are selected* and (2) the claim that evolution is a process *in which creatures are selected for their adaptive traits*. We will argue that: Darwinism is committed to inferring (2) from (1); that this inference is invalid (in fact it's what philosophers call an 'intensional fallacy'); and that there is no way to repair the damage consonant with commitment to naturalism, which we take to be common ground. Getting clear on all this will be a main goal of the book.

Why, you may reasonably ask, hasn't this tangle of connections been remarked upon before? We think the answer is pretty clear: although there has been a long and rich discussion of issues to which intensional explanations give rise, it is found almost entirely in the philosophical literature, which is not one that plays a large part in the education of biologists. Likewise the other way around: very few philosophers are sufficiently conversant with the tradition of evolutionary theorizing in biology to understand how much it relies on an unexplicated notion of 'selection-for'. When philosophers have thought about intensional explanation, it has almost invariably been intensional psychological explanation that they have had in mind. It seems, in retrospect, that an extensive interdisciplinary discussion of evolutionary theory between philosophers and biologists might have proved profitable. But, of course, everybody is busy and you can't read everything. Nor can we.

There are other examples of issues about which useful interdisciplinary discussions of adaptationism might have occurred but did not. A recurrent theme in what follows is the important analogy between the account of the fixation of phenotypes that Darwin offered and the 'learning theoretic' account of the acquisition of 'behavioural repertoires' promoted by the once very influential Harvard psychologist Burrhus Frederic Skinner, a father of behaviourism. In fact, we claim that Skinner's account of learning and Darwin's account of evolution are identical in all but name. (Probably Skinner would have agreed with us; he made frequent attempts to shelter under Darwin's wing.) B. F. Skinner was perhaps the most notable academic psychologist in America in the mid-twentieth century. Certainly he was the most widely discussed. His explicit goal was to construct a rigorous and scientific account of how learned behaviours are acquired. The theory he endorsed blended the associationism of the British empiricists with the methodological positivism of psychologists such as Watson and philosophers such as Dewey. From the empiricists he inherited the thesis that learning is habit formation; from the positivists he inherited the thesis that scientific explanation must eschew the postulation of unobservables (including, notably, mental states and processes). Putting the two together produced a kind of psychology in which the organism is treated as a black box and learning is treated as the formation of associations between environmental stimuli and the behavioural responses that they elicit. The formation of such stimulus response associations was supposed to be governed by the law of effect – namely that reinforcement increases habit strength. These theses are, of course, a long way from any that Darwin held. But we'll see presently that what is wrong with Darwin's account of the evolution of phenotypes is very closely analogous to what is wrong with Skinner's account of the acquisition of learned behaviour.

Since the 1950s, it has been widely acknowledged that Skinner's project can't be carried out, and that the reasons that it can't are principled (for a classic review, see Chomsky, 1959; for later relevant arguments, see Chomsky's contributions in Piattelli-Palmarini, 1980). That being so, it is natural to wonder whether analogues of the objections that proved decisive against Skinner's learning theory might not apply, *mutatis mutandis*, against the theory of natural selection. In the event, however, evolutionary biologists do not read a lot about the history of behaviouristic learning theories, psychologists do not read a lot about evolutionary biology (although here the tide may be turning) and philosophers, by and large, do not read any of either. So, although the analogy between the theory of natural selection and the theory of operant conditioning has occasionally been remarked upon, the question of how the logic of the one might illuminate the logic of the other has rarely been seriously pursued. We hope to convince you that, once you've seen why Skinner can't have been right about the mechanisms of learning, it becomes pretty clear, for much the same reasons, that Darwin can't have been right about the mechanisms of evolution. Skinner was, of course, a behaviourist, and Darwin, of course, was not. But we will argue that the deepest problems that their theories face – versions of intensional fallacies in both cases – transcend this difference.

We've organized our discussion as follows.

Chapter 1 is about several ways in which learning theory and neo-Darwinist evolutionary theory are similar, both in their general architecture and in considerable detail. In particular, each is committed to a 'generate and filter' model of the phenomena it purports to explain; and each holds that, to a first approximation, the generator in question is random and the filter in question is exogenous. These assumptions have proved to be unsustainable in accounts of learning, and for reasons that would seem to apply equally to accounts of natural selection.

Part one is then devoted to recent research and thinking in biology. Chapters 2, 3 and 4 of Part one summarize a wealth of new facts and new non-selectional mechanisms that have been discovered in biology proper. They explain why our friends in biology are not 'that' kind of Darwinian any more. Chapter 5 offers a compendium of yet another kind of fact and explanation that's current in biology but alien to the standard neo-Darwinian theory of evolution. In essence, we report cases when optimal structures and processes have been found in biological systems. These are naturally occurring optimizations, probably originating in the laws of physics and chemistry. We think other self-organization processes by autocatalytic collective forces are almost sure to be elucidated in the near future. They are clearly, for reasons we detail in that chapter, not the outcome of natural selection winnowing randomly generated variations.

Part two then considers the logical and conceptual bases of the theory of natural selection. The 'cognitive science' approach to psychology that has largely replaced learning theory in the last several decades has stressed the role of endogenous constraints in shaping learned behavioural repertoires. *Pace* Skinner, what goes on in learning is not plausibly modelled as the exogenous filtering of behaviours that are in the first instance emitted at random. As we will have seen in Part one, there is a growing, very persuasive, body of empirical evidence that suggests something very similar in the case of the evolution of phenotypes: we think Darwinists, like Skinnerians, have overestimated the role of random generation and exogenous filtering in shaping phenotypes.

Chapters 6 and 7 are concerned with how issues of intensionality arise within adaptationist accounts of the mechanisms of natural selection. We start with the phenomenon of 'free-riding', in which neutral phenotypic traits are selected because they are linked with traits that causally affect fitness. Discussions of free-riding have become familiar in the biological literature since Gould and Lewontin (1979) (indeed, the phenomenon was recognized by Darwin himself), and it is widely agreed to constitute an exception to strictly adaptationist accounts of evolution. However, the consensus view is that it is a relatively marginal exception: one that can be acknowledged consistent with holding that the evolution of phenotypes is affected primarily by exogenous selection. We argue, however, that this consensus fails utterly to grasp the implications of free-riding and related phenomena for theories about how phenotypes evolve. Darwinists have a crux about free-riding because they haven't noticed the intensionality of selection-for and the like; and when it is brought to their attention, they haven't the slightest idea what to do about it.

We think that this situation has given rise to the plethora of spooks by which Darwinist accounts of evolution are increasingly haunted: Mother Nature, selfish genes, imperialistic memes and the like are the most familiar examples in the current literature. But the roots of free-rider problems go all the way back to Darwin's preoccupation with the putative analogy between the way that natural selection manipulates phenotypes and the way that breeders do. In this respect Darwin was inadequately impressed by the fact that breeders have minds – they act out of their beliefs, desires, intentions and so on – whereas, of course, nothing of that sort of is true in the case of natural selection. It would be startling, in light of this difference, if theories of the one could be reliable models for theories of the other.

Chapter 8 replies to what is perhaps the strongest argument in support of natural selection as the primary mechanism of the evolution of phenotypes: namely that no alternative can provide a naturalistic account of the 'exquisite adaptation' of creatures to their ecologies. We think this argument, although ubiquitous in the literature, is fallacious. This chapter explains why we think so.

In Chapter 9 we are interested in how issues about natural selection interact with more general questions about scientific explanation. Some empirical explanations characterize their domains simultaneously at several ontological levels. Historical explanations, for example, are egregiously 'multilevel'. An explanation of why Napoleon did what he did at Waterloo may advert simultaneously to his age, his upbringing, his social class and his personality type, to say nothing of his prior military experience, his psychological state, the weather and how much caffeine there was in his morning coffee. By contrast, there are 'single-level' theories – of which Newtonian mechanics is perhaps the extreme example. The individuals in the domain of that kind of theory are considered to consist entirely of point masses, and all the theory 'knows about them' (the parameters in terms of which the laws of the theory apply) are their locations and velocities and the forces acting upon them. Thus Newtonian mechanics abstracts from the colours of things, their individual histories, who (if anybody) owns them and so forth. It is often remarked that the individuals that are recognized by the basic sciences are more alike than the ones recognized by the special sciences. Whereas Newtonian particles differ only in their mass or their location or their velocity, organisms and phenotypes and ecologies can (and do) differ in all sorts of ways. See Ellis (2002) for an interesting discussion of this issue.

It is widely (but, we think, wrongly) supposed that evolutionary theories are basically of the single-level kind. The only relations they recognize are ones that hold among macro-level objects (and/ or events): organisms on the one side and their ecologies on the other. The explanations these sorts of theories offer specify how causal interactions between the organisms and the ecologies produce changes of fitness in the former. On this view, Darwin and Newton are in relatively similar lines of work.

Chapter 9 considers how plausible it is that evolution can be explained by a single-level theory. It is among the major morals this book wants to draw that, very possibly, there is nothing of interest that processes of phenotypic evolution have in common *as such*. If that is right, important things follow: as there is no single mechanism of the fixation of phenotypes, there is an important sense in which there is no 'level' of evolutionary explanation, and there can be no general theory of evolution. Rather, the story about the evolution of phenotypes belongs not to biology but to natural history; and history, natural or otherwise, is par excellence the locus of explanations that do not conform to the Newtonian paradigm.

We have also included an appendix of quotations that suggest how extensively hard-line versions of adaptationism have infiltrated fields adjacent to biology, including philosophy, psychology and semantics.

So much for a prospectus. We close these prefatory comments with a brief homily: we've been told by more than one of our colleagues that, even if Darwin was substantially wrong to claim that natural selection is the mechanism of evolution, nonetheless we shouldn't say so. Not, anyhow, in public. To do that is, however inadvertently, to align oneself with the Forces of Darkness, whose goal it is to bring Science into disrepute. Well, we don't agree. We think the way to discomfort the Forces of Darkness is to follow the arguments wherever they may lead, spreading such light as one can in the course of doing so. What makes the Forces of Darkness dark is that they aren't willing to do that. What makes Science scientific is that it is.

WHAT DARWIN
GOT WRONG

I

WHAT KIND OF THEORY IS THE THEORY OF NATURAL SELECTION?

Introduction

The (neo-)Darwinian theory of evolution (ET) has two distinct but related parts: there's a historical account of the genealogy of species (GS), and there's the theory of natural selection (NS). The main thesis of this book is that NS is irredeemably flawed. However, we have no quarrel to pick with the genealogy of species; it is perfectly possible – in fact, entirely likely – that GS is true even if NS is not. We are thus quite prepared to accept, at least for purposes of the discussion to follow, that most or all species are related by historical descent, perhaps by descent from a common primitive ancestor; and that, as a rule of thumb, the more similar the phenotypes of two species are,[1] the less remote is the nearest ancestor that they have in common.[2]

However, although we take it that GS and NS are independent, we do not suppose that they are unconnected. Think of the GS as a tree (or perhaps a bush) that is composed of nodes and paths; each node represents a species, and each species is an ancestor of whatever nodes trace back to it. The questions now arise: How did the taxonomy of species get to be the way that it is? What determines which nodes there are and which paths there are between them? In particular, by what process does an ancestor species differentiate into its descendants? These are the questions that Darwin's adaptationism purports to answer. The answer it proposes is that if, in the genealogical tree, node A traces back to node B, then species B arose from species A by a process of *natural selection*, and the path between the nodes corresponds to the operation of that process.

We will argue that it is pretty clear that this answer is not right; whatever NS is, it cannot be the mechanism that generates the historical taxonomy of species. Jared Diamond in his introduction to Mayr (2001, p. x) remarks that Darwin didn't just present '... a well-thought-out theory of evolution. Most importantly, he also proposed a theory of causation, the theory of natural selection.' Well, if we're right, that's exactly what Darwin *did not* do; or, if you prefer, Darwin did propose a causal mechanism for the process of speciation, but he got it wrong.

There are certain historical ironies in this because it is the Darwinian genealogy, and *not* the theory of natural selection, that has been the subject of so much political and theological controversy over the last hundred years or so. To put it crudely, what people who do not like Darwinism have mostly objected to is the implication that there's a baboon in their family tree; more precisely, they do not admit to a (recent) ancestor that they and the baboon have in common. Accordingly, the question doesn't arise for them how the ancestral ape evolved into us on the one hand and baboons on the other. This book is anti-Darwinist, but (to repeat) it is not *that* kind of anti-Darwinist. It is quite prepared to swallow whole both the baboon and the ancestral ape, but not the thesis that NS is the mechanism of speciation.

The argument for the conclusion that there is something wrong with NS is actually quite straightforward; to some extent, it's even familiar. Not, however, from discussions of Darwinism per se, but from issues that arise in such adjacent fields as the metaphysics of reference, the status of biological teleology and, above all, in the psychology of learning. Bringing out the abstract similarity – indeed, identity – of this prima facie heterogeneous collection is a main goal in what follows. But doing so will require a somewhat idiosyncratic exposition of NS.

In the first place, we propose to introduce NS in a way that distinguishes between: (1) the theory considered simply as a 'black box' (that is, simply as a function that maps certain sorts of inputs onto certain sorts of outputs); and (2) the account that the theory gives of the mechanisms that compute that function and of the constraints under which the computations operate. This is, as we say, a somewhat eccentric way of cutting up the pie; but it will pay its way later on when we try to make clear what we take to be the trouble with NS.

In the second place, we want to develop our exposition of Darwin's account of evolution in parallel with an exposition of B. F. Skinner's theory of learning by operant conditioning (OT). Some of the similarities between the two have been widely noted, not least by Skinner himself.[3] But we think, even so, that the strength of the analogy between NS and OT has been seriously underestimated, and that its implications have generally been misunderstood. In fact, the two theories are virtually identical: they propose essentially the same mechanisms to compute essentially similar functions under essentially identical constraints. This raises a question about which prior discussions of NS have been, it seems to us, remarkably reticent: it is pretty generally agreed, these days, that the Skinnerian account of learning is dead beyond resuscitation. So, if it is true that Skinner's theory and Darwin's are variations on the same theme, why aren't the objections that are routinely raised against the former likewise raised against the latter? If nobody believes Skinner any more, why does everybody still believe Darwin? We're going to argue that the position that retains the second but not the first is not stable.

Natural selection considered as a black box

As just remarked, one way to think about NS is as an account of the process that connects ancestral species with their descendants. Another (compatible) way is to think of it is as explaining how the phenotypic properties of populations change over time in response to ecological variables.[4] By and large, contemporary discussions of evolution tend to stress the second construal; indeed, it's sometimes said that this sort of 'population thinking' was Darwin's most important contribution to biology.[5]

Whether or not that is so, 'population thinking' is convenient for our present purposes; it allows us to construe an evolutionary theory abstractly, as a black box in which the input specifies the distribution of phenotypes at a certain time (the G_N [Generation N] distribution, together with the relevant aspects of its ecology), and in which the output specifies the distribution of phenotypes in the next generation (G_{N+1}). This provides a perspective from which the analogies between NS and OT become visible, since OT is also plausibly viewed as a black box that maps a distribution of traits in a population at a time (a creature's behavioural repertoire at that time), together with a specification of relevant environmental variables (viz. the creature's history of reinforcement), onto a succeeding distribution of traits (viz. the creature's behavioural repertoire consequent to training). We therefore propose, in what follows, to indulge in a little 'population thinking' about both NS and OT.

Operant conditioning theory considered as a black box

If we are to think of the Skinnerian theory of learning in this way, we will first have to decide what is to count as a 'psychological trait'. Fortunately, Skinner has an explicit view about this, which, although by no means tenable, will serve quite nicely for the purposes of exposition.

Let's stipulate that a creature's 'psychological profile' at a certain time is the set of psychological traits of the creature at that time.[6] For Skinner, a psychological trait is paradigmatically a stimulus–response (S–R) association; that is, it is a disposition to perform a token of a certain type of behaviour 'in the presence of' a token of a certain type of environmental event.[7] Skinner takes S–R associations to be typically probabilistic, so a creature's psychological profile at a certain time is a distribution of probabilities over a bundle of S–R associations. Correspondingly, OT is a theory about how the distribution of probabilities in a population of S–R connections varies over time as a function of specified environmental variables (including, notably, 'histories of reinforcement'). The picture is, in effect, that the totality of a creature's dispositions to produce responses to stimuli constitutes its psychological profile. These dispositions compete for strength, and environmental variables determine which dispositions win the competitions; they do so in accordance with the laws of conditioning that OT proposes to specify, and of which the so-called 'law of effect' is the paradigm.

We've been describing OT as a kind of 'population thinking' in order to emphasize its similarity to evolutionary theory (ET): both are about how traits in a population change over time in response to environmental variables (*ecological* variables; see footnote 5). That is, we suppose, a mildly interesting way of looking at things, but if it were all that the ET/OT analogy amounted to, it would warrant only cursory attention. In fact, however, there is quite a lot more to be said. Both theories postulate certain strong constraints (we'll call them 'proprietary' constraints)[8] on how the empirical facts about population-to-population mappings are to be explained; and in both cases, the choice among candidate theories relies heavily on the imposition of these constraints.[9] Some proprietary constraints derive from (what purport to be) general methodological considerations;[10] but many of them are contingent and substantive. They derive from assumptions about the nature of evolution on the one hand and of learning on the other. The substance of the analogy between Darwin's version of evolutionary theory and Skinner's version of learning theory consists, in part, in the fact that the proprietary constraints that they endorse are virtually identical.

Proprietary constraints (1): iterativity

OT and NS are both formulated so as to apply 'iteratively' in their respective domains. That's to say that psychological profiles are themselves susceptible to further conditioning, and evolved phenotypes are themselves susceptible to further evolution. Iterativity is required in order that OT and ET should acknowledge the open-endedness of their respective domains: ET implies no bounds on the varieties of phenotypes that may be subject to evolution, and OT implies no bounds on the variety of behavioural profiles that may be modified by learning. The effect of this is to permit both theories to begin their explanations *in medias res*. ET presupposes some presumably very simple unevolved self-replicators with phenotypic traits to which the laws of evolution apply in the first instance; OT presupposes some presumably very simple repertoire of S–R associations to which the putative laws of conditioning apply in the first instance. In both cases, there are serious questions as to exactly what such 'starting assumptions' a theorist ought to endorse. In OT, the usual view is that an organism at birth (or perhaps *in utero*) is a random source of behaviours. That is, prior to operant learning, any stimulus may evoke any response, although the initial probability that a given stimulus will evoke a given response is generally very small. In ET, a lot depends on what kind of self-replicator evolutionary processes are supposed to have first applied to. Whatever it was, if it was *ipso facto* subject to evolution, it must have been a generator of heritable phenotypes, some of which were more fit than others in the environmental conditions that obtained.

Proprietary constraints (2): environmentalism

What phenotypes there can be is presumably determined by (among other things) what genotypes there can be; and what is genotypically possible is constrained by what is possible at 'lower' levels of organization: physiological, genetic, biochemical or whatever. Likewise for the effects of physiological (and particularly neurological) variables on psychological phenomena. It is, however, characteristic of both ET and OT largely to abstract from the effects of such endogenous variables, claiming that the phenomena of evolution on the one hand and of psychology on the other are very largely the effects of environmental causes.

A striking consequence of this assumption is that, to a first approximation, the laws of psychology and of evolution are both supposed to hold very broadly across the phylogenetic continuum, abstracting both from differences among individuals and from differences among species. (In the darkest days of conditioning theory, one psychologist claimed that, if we had a really adequate theory of learning, we could use it to teach English to worms. Happily, however, he later recovered.) Likewise, it is characteristic of evolutionary biologists to claim that the same laws of selection that shape the phenotypes of relatively simple creatures such as protozoa also shape the phenotypes of very complex creatures such as primates. It's clearly an empirical issue whether, or to what extent, such environmentalist claims are true in either case. It turned out that OT greatly underestimated the role of endogenous structures in psychological explanation; much of the 'cognitive science' approach to psychology has been an attempt to develop alternatives to OT's radical environmentalism. In Part one we will consider a number of recent findings in biology that suggest that analogous revisions may be required in the case of ET.[11]

Proprietary constraints (3): gradualism

ET purports to specify causal laws that govern transitions from the census of phenotypes in an ancestral population to the census of phenotypes in its successor generation. Likewise, OT purports to specify causal laws that connect a creature's psychological profile at a given time with its succeeding psychological profile. In principle, it is perfectly possible that such laws might tolerate radical discontinuities between successive stages; gradualism amounts to the empirical claim that, as a matter of fact, they do not. This implies, in the case of ET, that even speciation is a process in which phenotypes alter gradually, in response to selection pressure.[12] 'Saltations' (large jumps from a phenotype to its immediate successor) perhaps occur from time to time; but they are held to be sufficiently infrequent that theories of evolution can generally ignore them.[13] In OT, gradualism implies that learning curves are generally smooth functions of histories of reinforcement. Learning consists of a gradual increment of the strength of S–R associations and not, for example, in sudden insights into the character of environmental contingencies. Strictly speaking, according to OT, there is no such thing as problem solving; there is only the gradual accommodation of a creature's behaviours and behavioural dispositions to regularities in its environment.

In neither learning nor evolution is the claim for gradualism self-evidently true. Apparent discontinuities in the fossil record were a cause of considerable worry to Darwin himself, and there continues to be a tug-of-war about how they ought to be interpreted: evolutionary biologists may see fortuitous geological artefacts where palaeontologists see bona fide evidence that evolution sometimes proceeds in jumps (Eldredge, 1996). Likewise, a still robust tradition in developmental psychology postulates a more-or-less fixed sequence of cognitive 'stages', each with its distinctive modes of conceptualization and correspondingly distinctive capacities for problem solving. Piagetian psychology is the paradigm; for decades Piaget and Skinner seemed to be exclusive and exhaustive approaches to the psychology of learning.[14]

It is thus possible to wonder why gradualism has seemed, and continues to seem, so attractive to both evolutionary theorists and learning theorists. Some of the answer will become apparent when we, as it were, open the two black boxes and consider how ET and OT go about computing their respective outputs. Suffice to say, in the meantime, that the case for evolutionary gradualism was strengthened by the 'modern synthesis' of evolutionary biology with genetics. To a first approximation, the current view is that alterations of phenotypes typically express corresponding alterations of genotypes, alterations of genotypes are typically the consequence of genetic mutation, and macromutations generally decrease fitness. If all that is true, and if evolution is a process in which fitness generally increases over time, it follows that saltations cannot play a major role in evolutionary processes.

The allegiance to gradualism in the psychology of learning is perhaps less easily explained; at a minimum, there would appear to be abundant anecdotal evidence for discontinuities in cognitive processes that mediate learning, problem solving and the like ('and then it suddenly occurred to me ...', 'and then we realized ...' and so forth).

But it's important to bear in mind that OT is a direct descendant of the associationism of the British empiricists. In particular, OT inherited the empiricist's assumption that learning consists mostly of habit formation; and, practically by definition, habits are traits that are acquired gradually as a consequence of practice. In this respect, the differences between Skinner and (e.g.) Hume turn mostly on issues about behaviourism, not on their theories of learning per se. Both think that learning is primarily associative and that association is primarily the formation of habits.

Proprietary constraints (4): monotonicity

If the ecology remains constant, selection increases fitness more or less monotonically;[15] likewise for the effects of operant condition in increasing the efficiency of psychological profiles.[16] The motivation for these constraints is relatively transparent: ET and OT are one-factor theories of their respective domains. According to the former, selection is overwhelmingly decisive in shaping the evolution of phenotypes;[17] according to the latter, reward is overwhelmingly decisive in determining the constitution of psychological profiles. Because, by assumption, there are no variables that interact significantly with either selection or reinforcement, the monotonicity of each is assured: if selection for a phenotypic trait increases fitness on one occasion, then it ought also to increase fitness on the next; if a certain reinforcing stimulus increases the strength of a certain response habit, the next reinforcement should do so too.[18]

These are, of course, very strong claims. In real life (that is, absent radical idealization) practically nothing is a monotonic function of practically anything else. So perhaps it's unsurprising that there are counter-intuitive consequences, both for theories of evolution, according to which the effect of selection on fitness is monotonic, and for theories of learning, which claim monotonicity for the effects of reinforcement on habit strength. What is interesting for our present purposes, however, is that the prima facie anomalies are very similar in the two cases. Thus, for example, OT has notorious problems with explaining how 'local maximums' of efficiency are ever avoided in the course of learning, and ET has exactly the same problems with explaining how they are ever avoided in the course of selection. According to OT, creatures should persist in a relatively stupid habit so long as it elicits significant reinforcement; and that will be so even though there are, just down the road, alternative behavioural options that would increase the likelihood of reinforcement. Likewise, according to ET, if evolution finds a phenotypic trait that increases fitness, then selection will continue to favour that trait so long as the ecology isn't altered. This is so even if the phenotype that evolution has settled on is less good than alternative solutions would be. It is thus often said that evolution, as ET understands it, 'satisfices' but does not optimize: given enough time and a constant ecology, natural selection is guaranteed to converge on some fit phenotype or other; but if it happens to converge on the *best* of the possible adaptations, that's merely fortuitous. Exactly likewise in the case of the selection of S–R pairs by reinforcers. Neither ET nor OT provides a way of taking one step backwards in order to then take two steps forward.[19]

Plainly, however, the claim that evolution is a (mere) satisficer is prima facie a good deal more plausible than the corresponding claim about learning. Intuitively (though not, of course, according to OT), Scrooge can think to himself: 'I would be even richer if we didn't heat the office' and thence turn down the thermostat. But evolution can't think to itself 'frogs would catch still more flies if they had longer tongues' and thence lengthen the frog's tongue in order that they should do so.

Proprietary constraints (5): locality

The problems about local maximums exhibit one of a number of respects in which selection, as ET understands it, and learning, as OT understands it, are both 'local' processes: their operation is insensitive to the outcomes of merely hypothetical contingencies. What happened can affect learning or evolution; what might have happened but didn't *ipso facto* can't.

Likewise (and for much the same reasons) natural selection and reinforcement learning are insensitive to future outcomes (that the river will dry up *next week* does not affect any creature's fitness *now*; that the schedule of reinforcement will be changed on Tuesday does not affect the strength of any habits on Monday). Similarly for past events (unless they leave present traces); similarly for events that are merely probable (or merely improbable); similarly for events that happen too far away to affect the causal interactions that a creature is involved in; similarly for events from which the creature is mechanically isolated (there's an ocean between it and what would otherwise be its predators, and neither can fly or swim); similarly, indeed, for events from which a creature is causally isolated in any way at all. The general principle is straightforward: according to ET, nothing can affect selection except actual causal transactions between a creature and its actual ecology. According to OT, nothing can affect learning except actual reinforcements of a creature's actual behaviours.

We make a similar point to one we made above: although ET and OT have acknowledged much the same proprietary constraints, there is no principled reason why both of them should do so: it seems perfectly possible, for example, that selection should be locally caused even if learning is not. After all, learning, but not evolving, typically goes on in creatures that have *minds*, and minds are notoriously the kind of thing that may register the effects of events that are in the past (but are remembered) and of events that are in the future (but are anticipated) and events that are merely possible (but are contemplated) and so forth. We think, in fact, that whereas selection processes are *ipso facto* local, psychological processes are quite typically not. If we are right to think that, then the similarity of standard versions of ET and OT is a reason for believing that at least one of them is false.

Proprietary constraints (6): mindlessness

There is at least one way (or perhaps we should say, there is at least one sense) in which a creature can be affected by an event from which it is causally isolated: namely, it can be affected by the event *as mentally represented*. Thus: we consider cheating on our income tax; we find that we are very strongly tempted. 'But,' we think to ourselves, 'if we cheat, they are likely to catch us; and if they catch us, they are likely to put us in jail; and if they were to put us in jail, our cats would miss us'. So we don't cheat (anyhow, we don't cheat much). What's striking about this scenario is that what we do, or refrain from doing, is the effect of *how we think about things*, not of how the things we think about actually are. We don't cheat but we *consider* doing so; we don't go to jail, although the possibility that we might conditions our behaviour.

Now, causal interactions with events that are (merely) mentally represented would, of course, violate the locality constraint; it is presumably common ground that nothing counts as local unless it exists in the actual world. But mental representations themselves can act as causes, as when we cheat, or don't, because we've thought through the likely consequences. Darwin, however, held that the scientific story about how phenotypes evolve could *dispense with* appeals to mental causes. Indeed, one might plausibly claim that getting mental causes out of the story about how phenotypes evolve was his primary ambition.

There is, according to Darwin, no point at which an acceptable evolutionary explanation could take the form: such and such a creature has such and such a trait because God (or Mother Nature, or selfish genes or the Tooth Fairy) wished (intended, hoped, decided, preferred, etc.) that it should. This is so not only because there isn't any Tooth Fairy (and mere fictions do not cause things), but also because natural selection does not involve *agency*. That is, of course, a crucial respect in which the way *natural* selection is unlike *artificial* selection. If there are rust-resistant plants, that's because somebody decided to breed for them. But nobody decided to breed for the rust; *not even God*.[20] Mental causation (in particular, what philosophers call 'intentional causation')[21] literally does not come into natural selection; Skinner himself rightly emphasized that Darwin was committed to this;[22] not, however, because Darwin was a behaviourist (he wasn't) but because Darwin didn't believe in the Tooth Fairy (or, quite likely, in God either). There is, as we will presently find reason to emphasize, an occasional tendency among neo-Darwinians to flout this principle. That is entirely deplorable, and has caused endless confusion both in the journals and in the press.

We assume that Darwin was right that natural selection is not a kind of mental causation. It is not, however, at all obvious that the psychology of learning (or the psychology of anything else) can operate under a corresponding mindlessness constraint. There probably isn't a God or a Tooth Fairy; but there are minds, and they do have causal powers, and it is not implausible that one of their functions is to represent how things *might have been*, or *might be*, or *are in some other part of the forest*, or *would have been but that* ... and so forth. Commonsense psychology embraces causation by mental representations as a matter of course. By contrast, it's of the essence of Skinner's behaviourism, hence of OT, to deny that there is any such thing. In this respect, OT really was (and really was intended to be) a radical departure from commonsense ways of thinking about the mental. The analogy between OT and ET is exact in this respect: both prescind from the postulation of mental causes. The difference is that Darwin was right: evolution really is mindless. But Skinner was wrong: learning is not.

Thinking inside the boxes

So much for some of the similarities between ET and OT that emerge when they are viewed from the 'outside' – that's to say, from the perspective of what they propose to do rather than that of the mechanisms by which they propose to do it. Both are functions from states of populations to their successor states; and there are a number of substantive and methodological constraints that both endorse and are, to varying degrees, contentious. We now wish to change the point of view and consider the mechanisms that are proposed to implement these functions.

Two things strike the eye when the boxes are opened: first, the extremely exiguous character of the resources on which ET and OT rely to account for the rich and complex domains of data to which they are respectively responsible; and second, the all-but-identity of the causal mechanisms that the two theories postulate. We want to have a look at both of these. In a nutshell, ET and OT both offer 'generate and test' theories of the data that they seek to explain: each consists of a random generator of traits and a filter over the traits that are so generated. *And that is all.*[23]

We will look first at OT. It is convenient to do so because Darwin is, in certain crucial respects, less explicit about what mechanisms he thinks mediate the evolution of new phenotypes than Skinner is about what mechanisms he thinks mediate the fixation of new psychological profiles.

The mechanism of learning according to operant conditioning theory

It is crucial for Skinner that, in its initial state (which is to say, in abstraction from effects of prior learning), an organism is 'a random generator of operants.'[24] That is (prior learning and unconditioned reflexes aside), the psychological profiles on which OT operates are an unsystematic collections of S–R dispositions, each with an associative strength at or near zero. As previously remarked, OT undertakes to explain how reinforcement alters the strength of such dispositions in the direction of generally increasing efficiency.[25]

If, in the first instance, creatures generate S–R associations at random, then some 'shaping' mechanism must determine that the relative associative strength of some such pairs increases over time and that of others declines. It is characteristic of OT (as opposed, for example, to other varieties of associationism) to claim that shaping mechanisms are sensitive *solely* to exogenous variables; which is to say that association is sensitive solely to the effects of reinforcements on habit strength[26] in accordance with such laws of operant conditioning as, for example, the 'law of effect' (the strength of an association increases with the frequency with which it is 'followed by' reinforcement). In brief: S–R associations that are generated at random are then 'filtered' by a mechanism that implements the laws of association.

A word about 'random' generators: as gradualism is in force, the successor of a psychological profile cannot differ *arbitrarily* from its immediate ancestor. Reinforcement cannot, in one step, replace a low-strength S–R habit by a high-strength habit that connects some quite different stimulus to some quite different response. Perhaps reinforcing a random bar press in the presence of a light will produce a more urgent bar press next time the light goes on. But it won't produce a high-strength association between, as it might be, the light and an ear twitch; or between the bar press and the sound of a piano.[27] Reinforcement can lead to the association of 'new' kinds of responses (and/or of responses to new kinds of stimuli) but only via intermediate psychological profiles. The glaring analogy is to 'no saltation' theories of evolution (including ET), according to which the radical discontinuities between a creature's phenotype and the phenotype of its relatively remote ancestors must be mediated by the evolution of *intermediate phenotypic forms*. Accordingly, just as much of the serious scientific debate about OT has turned on whether learning curves are generally smooth enough to sustain its predictions, so much of the serious scientific debate about ET has turned on the extent to which the palaeontological record sustains the existence of intermediate forms in evolution.

The mechanism of selection according to evolutionary theory

According to the versions of Darwinism that have been standard since the 'new synthesis' of evolutionary theory with genetics, the overall picture is as follows.

(1) Phenotypic variation 'expresses' genotypic variation.[28]
(2) Genotypic variation from one generation to the next is the effect of random mutation.
(3) Macromutations are generally lethal.

(4) The phenotypic expression of viable mutations is generally random variation around population means.

In short, what ET says about the role of random genetic variation in the genesis of new species is exactly what OT says about the role of random operants in the formation of new behaviour profiles. The only relevant difference between the two is that random genetic variations can be heritable but random variations in the strength of operants cannot.

If the distribution of traits in a population is produced by filtering the output of a random generator, *what is the filter?* It's here that Skinner's story about the effect of conditioning in filtering randomly generated psychological profiles is more explicit than Darwin's story about the effect of selection in filtering randomly generated phenotypes. We will argue that, in fact, ET can offer no remotely plausible account of how filtering by natural selection might work. So here, finally, the analogy between OT and ET breaks down.

The putative laws of association provide Skinner with an account of how exogenous variables (in particular, schedules of reinforcement) filter populations of psychological profiles; they explain why the effect of such variables is that the relative strength of some habits increases over time and the relative strength of others does not.

So that answers the rhetorical question that is the title of this chapter. What kind of theory is Darwin's theory of natural selection? The same kind as Skinner's theory of operant conditioning. With, however, the following caveat: all that's wrong with Skinner's story about the filtering of psychological profiles is that it is a variety of associationism, and quite generally, associationism is not true. But Darwinism has (we'll claim) no analogous story about the evolutionary filtering of randomly generated phenotypes. In consequence, whereas Skinner's theory of conditioning is *false*, Darwin's theory of selection is *empty*.

So, anyhow, we will argue in Part two.

PART ONE

THE BIOLOGICAL ARGUMENT

2

INTERNAL CONSTRAINTS: WHAT THE NEW BIOLOGY TELLS US

> One can spend an entire lifetime correcting a flawed paper published in a reputable journal and still lose the battle if people like the basic idea.
>
> V. Hamburger, developmental neurobiologist, cited in Rakic, 2008

As we mentioned earlier, some of our good friends, patented experimental biologists (usually known as 'wet' biologists) who have read previous versions of this manuscript, slapped us on the wrist because they think what we are saying is overkill. They told us, 'no one is *that* kind of Darwinian any more'. We'd be happy if that were so, but there is good reason to doubt that it is. And, if it is true, the news has not been widely disseminated even among wet biologists (see, for example, Coyne, 2009).[1] This chapter and the next two are essentially a summary of *why* those biologists say what they (rightly) say. Chapter 5 wades into relatively new territory, even for biologists. News of what we summarize there has, alas, remained even more elusive so far.

Strict neo-Darwinists are, of course, environmentalists by definition: the genotype generates candidate phenotypes more or less at random; the environment filters for traits that are fitness enhancing.[2] But there are signs of a deep revisionism emerging in current evolutionary theory: modern biology urges us to conclude (what Darwin himself had acknowledged) that the effect of ecological variables on phenotypes is not the whole story about evolution. Indeed it goes further, urging us to conclude that ecological variables aren't even the most important part of the story about evolution. We will now see, in summary, how and why contemporary biology has changed classical neo-Darwinian adaptationism beyond recognition. Many important discoveries and many explicit quotes by their discoverers bear witness to this momentous change. Our book as a whole, however, parts company with many of these distinguished biologists. Paraphrasing a famous slogan by Karl Marx (an author whose views we do not consider to be otherwise germane), we can say: biologists have changed neo-Darwinism in many ways; the point now is to subvert it.

Natural selection is real, of course (when properly construed)

There can be little doubt that shifting equilibria[3] (that is, variations in the relative frequencies of phenotypic types within and across populations) happen all the time, on land, in the seas, in lakes, in rivers and in streams all over this planet. They also happen within our bodies. Alterations in epithelial (skin) cells, pancreatic cells, lymphocytes (white blood cells), neurons and synapses occurred in us even as we wrote these lines and in you even as you read them. Such shifts are relentless and have been happening on Earth for hundreds of millions of years. And webs of relations of predation, commensalism (food-sharing), competition and migration are intermingled with these shifts and modify, in the long run, our structure and that of our ecosystems. The distributions of biological and behavioural traits in populations that we see today are results of these processes, although certainly not exclusively so, and probably not even chiefly so (assuming that a reliable measure [a reasonable metric] could be established for such probabilistic evaluations, a topic to which we will return).

It's common ground that distributions of phenotypic traits in populations change slightly and relentlessly over time. Having said this much, however, it must be emphasized that such shifting equilibria do not explain the distribution of phenotypes; rather, they are among the phenomena that theories of evolution are supposed to explain. These days biologists have good reasons to believe that selection among randomly generated minor variants of phenotypic traits falls radically short of explaining the appearance of new forms of life. Assuming that evolution occurs over very, very long periods does not help if, as we believe, endogenous factors and multilevel genetic regulations play an essential role in determining the phenotypic options among which environmental variables can choose. Contrary to traditional opinion, it needs to be emphasized that natural selection among traits generated at random cannot by itself be the basic principle of evolution. Rather there must be strong, often decisive, endogenous constraints and hosts of regulations on the phenotypic options that exogenous selection operates on. We think of natural selection as tuning the piano, not as composing the melodies. That's our story, and we think it's the story that modern biology tells when it's properly construed. We will stick to it throughout what follows.

We think (and will argue in later chapters) that there are convincing a priori arguments that show this. For the moment, however, concede that it's often very hard to anticipate the effects of applying a process of selection to a randomly generated population of traits. Even slight variations in the initial frequencies, in the rates of random mutation and in the selection coefficients can lead to drastically different new equilibria.[4] This chapter summarizes a panorama of specific mechanisms the discovery of which makes the gradualist/adaptationist theory of natural selection plainly wrong in at least some cases, because new phenotypic traits aren't generated at random (as they would be if the mutations that they express are independent) or because adaptation to the ecology plays only a secondary role in the fixation of the phenotypes, or for both of these reasons.

Unidimensionality

The traditional doctrine of natural selection (NS) is unidimensional. Ecological structure is taken to be the explanation par excellence of phenotypic structure; the contribution of internal (endogenous) sources of variance and of internal constraints is, at most, deemed to be marginal. This suggests that for purposes of evolutionary explanation, one could abstract from the character of connections among genes and their phenotypic expressions, and also from the character of the genome itself. All of that internal structure is construed as largely irrelevant to explaining the course of evolution: NS will find its solutions regardless of genetic details. The clearest and most authoritative example of this sort of claim is to be found in Ernst Mayr, one of the main architects of the 'modern synthesis' (that is, the fusion of classical Darwinism with genetics, beginning in the first decades of the twentieth century). We quote what seems, in hindsight, and against the background of the 'evo-devo' revolution (see later in this chapter), a rather stunning statement:

> Much that has been learned about gene physiology makes it evident that the search for homologous genes is quite futile except in very close relatives. If there is only one efficient solution for a certain functional demand, very different gene complexes will come up with the same solution, no matter how different the pathway by which it is achieved. The saying 'many roads lead to Rome' is as true in evolution as in daily affairs.
>
> Mayr, 1963, p. 609

As we will see shortly, both the frequent cases of conservation of the same master genes across hundreds of million of years and the entire field of evo-devo tell a quite different story. The evo-devo revolution tells us that nothing in evolution makes sense except in the light of developmental biology – Mayr's statement is symptomatic of the unbounded power attributed to NS. A paradigmatic case was the formation of the eye across distant species, supposed to have taken place convergently and independently many times in evolution (at least five,

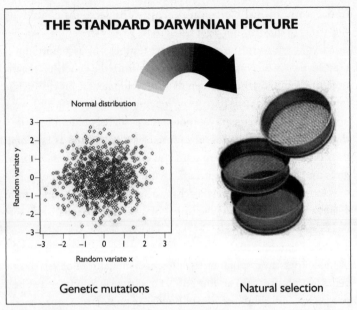

A schematic representation of the standard neo-Darwinian model of evolution by natural selection. The square on the left represents random genetic mutations, the arrow the expression of those mutations as manifest traits, (phenotypes) and the filters the action of natural selection.

maybe many more). But then, with the discovery of the same master genes for eye development (notably *Pax3*, *Pax2*, *Pax6* and *Dach*) across very distant classes and species (from the sea urchin, in which the genes remain unexpressed, to medusae, to fruit flies, to vertebrates) the evolutionary scene changed quite considerably (for stunning data and considerations, see Sherman, 2007).

As we will see in subsequent chapters, the picture that we get from the modern synthesis is like the one that behaviourist theory of learning proposes in psychology: a random generator of diversity, mapping onto phenotypes that, in turn, meet the demands imposed by environmental filters. The schema is something like that shown here.

Although they were no doubt always considered to be important in the world of real biological systems, considerations concerning the arrow in the diagram were taken to be largely irrelevant to evolutionary theory. As the quote from Mayr testifies, the assumption was that the theory of NS could abstract away from the details of genetic organization and from the details of developmental processes; the latter were deemed to be the concern of embryologists. Basically, in the new synthesis, convergent evolution was considered ubiquitous, rather than occasional, and endogenous variables were treated as random;[5] the exogenous variables were supposed do all the work. That is what gives the theory its unidimensional character. One 'big' arrow only.

Clear, indeed graphic, evidence of this way of thinking comes from some 80 years of picturing the course of evolution in terms of 'fitness landscapes' and 'adaptive landscapes'. The very first such graphs looked a lot like orographic contour maps, with valleys of low fitness and peaks of high fitness corresponding to different combinations of variants of genes (Wright, 1932). The effect of NS over time was supposed to be a progressive 'hill climbing' [sic] of biological populations, one generation after the other, up towards the peaks. Nice as they undoubtedly look graphically, there are many conceptual and practical problems with such maps. Frequently accompanied by mathematical equations, suggesting great scientific rigour, they remain nonetheless mere visual metaphors, as Pigliucci and Kaplan (2006) rightly emphasize.[6] The literature on this topic is huge, and we will not enter any details here.[7]

It must be noticed, at least in passing, that to assume the existence of a single, continuous, uni-valued mapping between gene configurations and the overall fitness of the organism neglects important factors, such as: the complexity of developmental pathways that are variable in systematic ways and constitute one of the many sources of internal constraints; the role of genomic imprinting (Peters and Robson, 2008) and of epigenetic factors (see below); and the impact of developmental noise (in the sense of Lewontin, 2000, Chapter 1) (developmental noise is a term that covers random microscopic events occurring at all levels, from individual cells to tissues, thus making even identical twins not completely identical, even at birth) (Fraga et al., 2005; Stromswold, 2006; Kaminsky et al., 2009). These internal fluctuations affect the course of phenotypic evolution prior to, and independent of, the effects of ecological variables.

Beanbag genetics

The range of new phenotypic options that are open at a given stage of evolution is, as we are going to see, drastically limited by internal constraints. Moreover, exogenous selection hardly ever operates on mutually independent traits. The idea that phenotypic traits can be independently selected was colourfully labelled 'beanbag genetics' by Ernst Mayr, who (to his credit) didn't believe it.

> The Mendelian was apt to compare the genetic contents of a population to a bag full of colored beans. Mutation was the exchange of one kind of bean for another. This conceptualization has been referred to as 'beanbag genetics'. Work in population and developmental genetics has shown, however, that the thinking of beanbag genetics is in many ways quite misleading. To consider genes as independent units is meaningless from the physiological as well as the evolutionary viewpoint.
>
> Mayr, 1963, p. 263

One of the founders of population genetics, J. B. S. Haldane, replied to Mayr, defending this approach in a classic 1964 paper.[8] Another British geneticist, Gabriel Dover, aptly tells us:

> It is naïve to assume that there are independent genes for each and every characteristic that have accumulated through past episodes of natural selection. The nature of biology is such that the basis of individuality is largely uncapturable, making all talk of the evolutionary origins of the unknown premature at best and vacuous at worst ... Selection is not a process as such with predictable outcomes based on fixed, selective 'powers' of individual genes controlling aspects of phenotype. Selection involves whole phenotypes, which are in part influenced by their unique combination of

genetic interactions; hence, evolution involves descent with modification of genetic interactions. Genetic interactions are phenotypic, not genotypic. Advocacy of the gene as the unit of selection is operationally incoherent and genetically misconceived.

Dover, 2006

A further crucial factor that militates against the idea of 'beanbag genetics' – that is, against the idea that inheritable variations in one trait are independent from inheritable variations in any other trait – lies in the convoluted 'packing' of genes in the chromosomes. The long 'strings' of DNA that form the genetic material are tightly coiled in the chromosomes in such a way that genes that are 'distant' in a purely sequential (linear) DNA ordering are brought spatially close and become, thus, susceptible to being jointly regulated. These intricate topological configurations of DNA make the joint regulation of gene expression within the same chromosome a rule, and even joint regulation of genes across different chromosomes (called 'kissing chromosomes') has been discovered (Kioussis, 2005). Moreover, in the cells of higher organisms, many proteins implicated in DNA repair cluster to form nuclear structures referred to as 'DNA repair factories', or 'foci'. The composition of these foci following different types of DNA lesions, the regulatory hierarchy of their assembly and the molecular details of events occurring inside these structures are under intense scrutiny (Meister *et al.*, 2003). As we will see, there are several units of regulation, spanning several genes, and these units are conserved not only across the successive cell divisions within a single organism, but often also across different species.

We will come back to this issue, but it's worth emphasizing right away that the assumption of atomistic (one trait at a time) mechanisms of natural selection is still at the core of many popular or semi-popular neo-Darwinian explanations. The structural 'solidarity' of several different traits, which have to be selected wholesale or not at all, makes 'free-riders' and accessory phenotypes not a rare exception, but rather the rule.[9] Surely, as a consequence of genetic, developmental and evolutionary modularity (see pages 48–9), in no living organism is everything effectively connected to everything else. If we imagine all the bits of structure and function in an organism as a very large pairwise interaction table, most of the cells in this table would be empty. Otherwise, evolution would have been impossible. The interesting question concerns the non-empty cells in that table: where they are and why. And the ensuing interesting question about evolution, in a genuinely modern perspective, is how it has taken place, given these local inter-dependencies, each one representing an evolutionary constraint.

Internal constraints and filters: 'evo-devo'

One common American phrase, sometimes said in a New England accent, is 'You can't get there from here'.[10] The 'there' in our case, is new theoretically possible species; the 'here' is an actual species, with all the constraints imposed by its internal structure. We saw above that the classical model of neo-Darwinism represented the manifest (phenotypic) consequences of internal changes in the genes (genotypic variants) as a unidimensional arrow from genotypes to phenotypes. In essence, it abstracted from all effects of development on visible traits, aside from the effects of genetic mutations, which were themselves considered to be largely independent of one another. But the internal developmental filters that neo-Darwinism tried so hard to abstract from now increasingly seem to be at the very core of evolution. Genes and phenotypes still count, of course;[11] but the evo-devo revolution[12] has stressed that evolution is essentially the evolution of the arrow that connects them. The slogan is: evolution is the evolution of ontogenies. In other words, the whole process of development, from the fertilized egg to the adult, modulates the phenotypic effects of genotypic changes, and thus 'filters' the phenotypic options that ecological variables ever have a chance to select from. The evo-devo revolution changes the classical picture quite considerably.[13] Clear statements to this effect are ubiquitous in the evo-devo literature:

> By viewing evolution as a branching tree of adults or genes,

theorists have omitted what selection really acts upon: ontogeny. Ontogenies evolve, not genes or adults. Mutated genes are passed on only to the extent that they promote survival of ontogenies; adulthood is only a fraction of ontogeny.

McKinney and Gittelman, 1995

In an evo-devo perspective, there is no reason for treating the first phase of a life cycle as if it was just a preparatory phase for the production of a living organism that will be sieved by the environment later on. From a formal standpoint, once production has occurred (i.e., when there are new individuals and new developmental processes have started, for instance from fertilized eggs), a downstream process that biases the composition of the 'bundle of ontogenetic trajectories' that constitutes a population is functionally a process of sorting, random as the lottery of life, nonrandom as natural selection, or a combination of the two.

Fusco, 2001

One foreseeable task of Evo-Devo is to set the limits between homology and bricolage, consisting of independent recruitment of a gene network and, in the end, to ascertain at what levels evolutionary constraints favour the recurrent invention of certain features, while preventing others from emerging.

Baguna and Garcia-Fernandez, 2003

The main discovery of evo-devo has been the remarkable invariance of the genetic building blocks of evolution. Because highly conserved master genes (see below, page 44 and *passim*) can persist through hundreds of millions of years of evolution, it is possible to perform experiments that exhibit aspects of genetic 'rescue'. This means that a 'healthy' variant of a given gene, if suitably inserted into the embryo at a very early stage and then activated, can successfully compensate for a 'defective' variant of that same gene (rescuing the function of the gene). This is an impressive biotechnological feat, but it stands to reason that it works. What's truly extraordinary is that such genetic rescue can *also* occur in organisms across distant species, illustrating both the extreme complexity of genotype to phenotype relations and the reality of the conservation of genes over evolutionary time. For instance, the standard, naturally occurring, version of a specific gene – the 'wild-type allele' in the technical vocabulary – from the fruit fly is able to 'rescue' a defective gene in the mouse, and vice versa. Further examples of the strict functional correspondence between genes across distant species are too numerous to mention.[14] New discoveries[15] of the deep similarity between genes in distant species, families, orders and even phyla continue to be published almost every month.[16]

Conservation of genes and gene complexes is not only compatible with variation in overall body plans typical of speciation, it is itself the principal source of such variation, by means of gene duplications, quadruplications and alternative switches in the regulation of these genes. Mirror images – binary polarity inversions – of very old gene complexes explain the difference, for instance, between the ventral organization of the nervous system in insects and its dorsal position in vertebrates.[17]

There are thus invariants in the developmental dynamics across evolutionarily distant animal forms, and there are also many genetic specificities that are preserved from phylum to phylum and from species to species. Moreover, comparisons of adults with adults, or of genomes with genomes, can be unrevealing when the units of trait transmission are whole pathways of development. *Pace* Ernst Mayr (see above), the identity of genes and gene complexes matters enormously in determining the process by which phenotypic properties can converge across different types of organisms: the conservation of genes and their roles in development over quite distant phyla and hundreds of millions of years of evolution is crucial to understanding such convergences. In the words of Nobel laureate Christiane Nüsslein-Volhard: 'this remarkable conservation [of genes and gene complexes] came as a great surprise. It had been neither predicted nor expected'. Certainly it's remarkable, given that phenotypic structure is presently modelled as largely the outcome of endogenous variables.

Some (undue) perplexities about the evo-devo revolution

The discovery that the same genes and gene complexes are found across very different forms of life, spanning hundreds of millions of years of evolution, has not failed to raise initial perplexity in the biological profession. In 2002, in a commentary/review of work in evo-devo up to that moment, Elisabeth Pennisi interviewed several (in her own words) 'evo-devo enthusiasts' in *Science* and reported their growing puzzlement, when they 'get down to details' (Pennisi, 2002). In the overall economy of this book, and of this chapter in particular, we think that this piece is very revealing and we indulge here in some verbatim quotes. Pennisi rightly states that evo-devo has turned a famous motto by the evolutionist Theodosius Dobzhansky on its ear. Dobzhansky, in a lay sermon to American teachers of biology in 1973, ventured to state that 'nothing in biology makes sense except in the light of evolution' (Dobzhansky, 1973). Evo-devo tells us that it's the other way around: nothing in evolution makes sense except in the light of developmental biology. But it has not been easy, and it still isn't, to set out in precise evolutionary terms what is known about genes and development. Researchers have been grappling for some years with the problem of reconstructing the way in which similar genes mastermind the development of wildly different creatures. William Jeffery, an evolutionary developmental biologist at the University of Maryland, College Park, told Pennisi: 'You can collect lists of conserved genes, but once you get those lists, it's very hard to get at the mechanisms [of evolution]'. His conclusion is rather drastic: 'Macroevolution is really at a dead end.' Jeffery's colleague at Maryland, Eric Haag (significantly, we think) adds that the fundamental question is whether the mutations that result in real novelty are the same mutations that happen day to day or are the ones that occur only rarely, on a geological timescale. Rudolf Raff, an evo-devo researcher at Bloomington, Indiana, told Pennisi that, since variation is the very stuff of evolution, 'what developmental biologists consider noise, the [micro]evolutionists consider gold.'

The data on the remarkable conservation of genes are, however, incontrovertible, and constantly growing in quantity and detail. In the last few years, a vast body of research and manipulations of various sorts on model organisms (such as the omnipresent fruit fly, the nematode [a roundworm], the tiny zebrafish, the chick embryo, and so on) has shed light onto a variety of subtle genetic regulations. The initial perplexities we have evoked above are being superseded by new concepts, new avenues of research and new models of the basic evolutionary dynamics. As we will see in a moment, multiple levels of regulation act on the expression of genes at various stages of development.

Summary on the lessons from evo-devo

The very least that can be said, in the light of evo-devo, is that a uni-dimensional theory of evolution hasn't a prayer of being adequate. The frequent conservation of genes and gene complexes refutes the idea that morphological and functional convergences are, almost always and everywhere, to be construed as adaptive 'solutions' to cor-respondingly ubiquitous survival 'problems'.[18] But we do not want, either, to be taken as committing the fallacy of suggesting that, if a theory cannot explain everything, then it cannot explain anything. For the sake of the argument, let's concede that there are some prima facie plausible cases of evolutionary convergence not explicable by common descent (see Rueber and Adams [2001] for examples con-cerning dentition, body shape and the shape of head and mouth of the cichlids [fish species] in Lake Tanganyika). And there are prima facie plausible cases of morphological and behavioural adaptation quite probably caused by environmental changes (as in the changes of blood density and the loss of haemoglobin in several species of icefish in the Antarctica – for a lay description of this case see Carroll [2006]. Yet, recent advances in evo-devo show that phenotypic conver-gence is, more often than not, the effect of genetic and developmental invariants. Conversely, numerous examples have been found, some replicated in the laboratory,[19] of remarkable differences in terminal forms produced by slight variations in the regulation of the same gene complexes and/or in the timing of activation of such complexes. The interesting consequence is that the huge variety of extant and fossil life forms ('endless forms most beautiful', in the words of Sean Carroll [2005], borrowed from Darwin), is not only fully compatible with the high conservation of genes, but also explained by it. They are, in other words, explained by the complex intermingling of genetic conservation and variable gene regulations, at various levels.[20] The actual cases of, respectively, evolutionary convergence in the cichlids and adaptation to extreme cold in the icefish do not lend themselves to being generalized as evolutionary standards, as offering a single layer of evolutionary explanations. (We will return in some detail in the following chapters to what we think is the true nature and import of this different kind of evolutionary explanation.)

Much more of much the same

Gene conservation and the conservation of developmental (technically called 'ontogenetic') processes are two of the ways that phenotypes can converge even as the corresponding ecologies vary. Neither of these claims is seriously disputed in current biological discussions; and both imply internal filters on the phenotypes on which exogenous selection acts. They challenge the classical neo-Darwinist view that the course of evolution is exhaustively driven by exogenous factors. The old argument in evolutionary biology was about whether internal constraints are the exceptions or the rule; the present consensus is increasingly that they are the rule. At a minimum, there are plenty of other examples of internal filters on phenotypic variables, and they are to be found at a variety of levels of endogenous structure. Here are some more.

Genetic mutation is basically a quantum phenomenon, chance substitutions of one of the four 'letters' (nucleotides) in a DNA sequence for another. Therefore, at their very source, mutations may occur at random; but their effects are not uniform either across the different positions inside the affected gene, or the position that gene occupies in the whole genome, or across species. For example, there are several known instances of regions in genomes that are called 'hypermutable' and there are 'hotspots'. What this means, as these labels clearly indicate, is that mutations are more likely to affect these genetic regions than others (Shen and Storb, 2004). Some of these regions are likely to produce tumours in humans and other species (Laken *et al.*, 1997). Slippage during the gene replication process, resulting in gains or losses of repeat DNA units, is a cause, but not the only one. The so-called minisatellites are especially affected (Yauk, 1998). These form a class of highly variable (polymorphic) 'tandem repeats' in the DNA sequence. They include some of the most variable spots in the human genome, with mutation rates ranging from 0.5 to over 20 per cent per generation (Bois, 2003). At the opposite extreme, various processes of repair (DNA repair) (Feuerhahn and Egly, 2008) act as a buffer against mutations (for recent findings in the human genome, see Berglund *et al.*, 2009; Hurst, 2009). Mutations in the proteins that execute this task are usually lethal.

In essence, thus, the traditional assumption that mutations have a fixed probability of occurring anywhere at random in the genome of any species (something like one chance in a million, per locus per generation) does not stand up to a more refined scrutiny. It surely is not the case that there are random corresponding effects downstream. In other words, even if they were actually random, mutations would not always produce random novel phenotypes.

In fact, the next stages of the processes that connect genomes with phenotypes reveal still other mechanisms that uncouple random mutations of the DNA from their phenotypic consequences. For example, the transcription of DNA into messenger RNA (mRNA), the very first process towards gene expression, has multiple internal regulators. RNA editing (the term suggests intuitively what it's all about) effectively alters the transcript of a gene into an mRNA, so that the chemical (amino acid) sequence of the actually encoded protein differs from the one that would be predicted by the original genomic DNA sequence. This central process, the transcription of whole genes, the step through which the DNA of the genes is transferred (technically, 'transcribed') into the daughter molecule called RNA, is subject to a manifold of regulatory processes. These contribute in crucial ways to determine which genes are activated when, in early development and then in the later life of the individual. Some such mechanisms are ubiquitous across all species; others are specific to particular species.[21]

The toast of the town has been, in recent years, the study of the so-called micro-RNAs (miRNAs), short non-coding sequences of RNA that have ubiquitous and crucial regulatory functions.[22] These micro-sequences of RNA, only about 20 to 25 genetic characters (bases) long, act on the much longer sequences of mRNA, the primary 'transcript' of the genetic material (DNA), that typically are many thousands of bases long. Since they regulate the transcription of DNA into RNA, the miRNAs regulate the expression of genes. In animals, miRNAs have hundreds of targets, which may include non-coding RNAs (Zhao *et al.*, 2003), and regulate development in a range of ways, for example by targeting genes in key signalling pathways.

The role of these mechanisms in evolution is only beginning to be unravelled (Filipowicz *et al.*, 2008). In a recent (2008) review, two leading experts, the Australian molecular geneticists Paulo P. Amaral and John S. Mattick, say:

> These [evo-devo and multiple regulation] perspectives, and the evidence that increasingly supports them, are at odds with the orthodox assumption that the vast majority of the mammalian genome is not functional and consequently the vast majority of the RNAs transcribed in the cells are not meaningful. On the contrary, we suggest that the mammalian genome, rather than being viewed as islands of protein-coding sequences in a sea of evolutionary junk, may be more accurately thought of as an RNA machine, wherein most information is expressed as non-coding (nc)RNAs in a developmentally regulated manner to orchestrate the precise patterns of gene expression during mammalian ontogeny. This emerging view does not conflict but will have to be reconciled and integrated with the well-described protein-based regulatory, signalling, and effector networks that are also central to multicellular development.
>
> Amaral and Mattick, 2008, p. 479

In a January 2009 editorial celebrating the 50th anniversary of the journal *Developmental Biology*, Editor-in-chief Robb Krumlauf said: 'A major challenge for the future will be to decipher how the basic gene "tool kit" and common signalling pathways are controlled and integrated in the development and evolution of so many distinct organisms.'

The list could be continued with RNAi (i stands for 'interference') and various processes of 'proofreading'. There are also processes of post-transcriptional silencing, adding a further mechanism of regulation.

Simplifying rather drastically,[23] then, the messenger RNA exits the cell nucleus and goes into the cell factories (the ribosomes) and is translated into proteins. Proteins are the primary stuff of which life is made, and they literally peel off the ribosomes, each folding into a specific spatial conformation, jointly determined by its chemical (amino acid) sequence and the medium in which it folds (water, lipids, etc.)[24] (Dobson, 2003). The three-dimensional spatial configuration of each protein determines its biological function and has to be attained quite exactly, or else ...

Chaperones

Yes, or else. But other proteins called 'chaperones', and an important subset thereof called 'chaperonines', secure what is called 'quality control', that is, the correct folding of their 'client' proteins. One of the best-known chaperones is called HSP90 (heat shock protein 90). It assists the folding of a huge variety of proteins. A mutation in this protein generates all sorts of monstrosities in the fruit fly. Some such monsters may be selectively bred in the laboratory and transmit their anomalous phenotype to the next generations. After several artificial selections, some individuals carrying these anomalies are produced and the anomalous traits are stably maintained, even after the HSP90 has been brought back, by genetic implants, to its normal state.[25] Geneticists these days can perform such marvels in their laboratories and can conclude that proteins such as HSP90 act as 'evolutionary capacitors' (Rutherford and Lindquist, 1998; Queltsch *et al.*, 2002;

True *et al.*, 2004). This means that some potentially deleterious mutations (they would be so, if they were expressed there and then) can be
kept at bay for generations, passed on from one generation to the next
but remaining inert, until some other mutation, or major changes in
the environment (shocks, see page 58), can expose them, that is, can
make of that genotype a corresponding phenotype.[26] A direct molecular interaction between HSP90 and a protein of the chromatin (called
Trithorax, or for short TRX) has been reported recently. Since the
chromatin proteins control the developmental fate of cells by modu-
lating epigenetic signals (see below), these data explain in detail the
central role of HSP90 in cooperating with these proteins in maintaining the active expression state of target genes, notably including
master genes such as the Hox genes. When HSP90 is damaged, either
by genetic mutations or by external pharmacological inhibition, the
function of these master genes is downregulated, with the ubiquitous
and dramatic consequences we have indicated above.

Alternative splicing

Finally (for the purposes of the present summary exposition) the
segments of a gene that actually encode segments of proteins (these
are called exons, while the segments that do not code for proteins
are called introns) can be spliced in different alternative ways.[27] As
a result, a single gene can code for many different proteins, and a
single mutation in one of the exons can affect many of these proteins in a single stroke. When the human genome was decoded there
was a lower than expected number of genes (of the order of only
24,000), prompting renewed interest in alternative splicing, as a way
for a single gene to encode many proteins. Genes were supposed to be
'multitasking'. And they are.[28]

A recent study by Christopher B. Burge of MIT and colleagues
analysed the entire sequence of the messenger RNA (mRNA) of
fifteen different human tissue types or cancer cell lines to produce
a comprehensive catalogue of gene and alternative mRNA expression (Wang *et al.*, 2008). Over 90 per cent of human genes are now

estimated to undergo alternative splicing, exploiting this form of mRNA processing that yields multiple proteins from a single gene. Almost all mammalian genes have interruptions in their coding regions and are alternatively spliced. This mechanism permits greater phenotypic complexity than indicated by gene number alone.[29]

Last but not least: molecular drive and biased gene conversion

Deploring the traditional monopoly attributed to natural selection and random genetic drift as the sole causes of the formation of new species, Gabriel Dover proposed a new mechanism, called 'molecular drive' (Dover, 1982a,b). In essence, Dover's proposal consisted of a molecular process of turnover internal to the genome, independent of natural selection. A concerted pattern of fixation that permits the formation of novel biological forms 'in a manner not predicted by the classical genetics of natural selection and genetic drift'. In explaining, with the tools available in the early 1980s, the possible details of such completely mechanistic molecular processes, totally internal to the genetic apparatus, Dover used three words that have become quite prominent, more than 25 years later: 'directional', 'biased' and 'conversion'.

In essence, Dover stresses that all genomes of all examined species from bugs to worms to humans are riddled with the ubiquitous genomic mechanisms of turnover (replicative transposition, inversion, duplication) that power what he calls molecular drive. The recurrent instability of genomes leads to reorganizations and to new temporary stabilizations. Dover stresses that the spreading consequences of molecular drive also work in exactly the same way (sampling error) that stochastic genetic drift works at the phenotype level. Besides gene conversion (biased and unbiased), to which we will return in a moment, there are transpositions, slippages, unequal crossing over of chromosomes and other processes, which together ensure, in Dover's picture, that what starts off as a single mutation in a single gene in a single chromosome in a single individual can, with the passing of the generations, spread throughout a sexually reproducing population. This internally driven spreading process can open up, for a population, in the course of time, paths of development and reproduction and behaviour that were previously inaccessible to it. In Dover's schema (see also his 2001 book, *Dear Mr Darwin*), the establishment of novel, environmentally friendly functions can be envisaged as dependent on an interaction between many processes: molecular drive, random drift and (yes, also) natural selection. In his schema, summarizing it drastically, there are forces at work that are basically due to the instability of genomes (ubiquitous non-Mendelian mechanisms of turnover). These provide a radically wider comprehension of the evolved nature of biological functions.

One aspect of Dover's earlier intuitions (and data, and calculations) is now being emphasized: the process called biased gene conversion (BGC).[30] This mechanism, related to gene recombination with ensuing segregation and distortion, is presently observed to drive the fixation of new gene variants (new alleles) independently of any selective process. A class of numerous and evolutionarily recent differences in DNA sequences between corresponding genes in humans and in non-human primates that were previously attributed to intense natural selection now appear to be due to BGC (Berglund *et al.*, 2009). The decisive impact of BGC on traditional conceptions (and statistical calculations) of alleged selective 'sweeps' in human evolution is also stressed in Duret, 2009; Galtier *et al.*, 2009; Hodgkinson *et al.*, 2009. How distinct these genetic conversion processes are from any semblance of natural selection is shown by the fact that they can even promote the fixation of deleterious mutations in primates.

The English biochemist Laurence Hurst, in a commentary in *Nature* on 29 January 2009, writes of the data produced by Berglund *et al.*:

> The[se] results ... accord with the view of BGC as a driver of sequence evolution, potentially explaining the occurrence of large spans of approximately homogeneous nucleotide content ... in our genome. More disturbingly, the results bring into question the usefulness of the standard tool kit for identifying hotspots of changes that are beneficial to organisms. Convincing demonstration of positive selection now requires both evidence that the

changes were not caused by BGC and scrutiny of the impact of the amino-acid changes.

Hurst, 2009, p. 544

Conclusion to this chapter

Perhaps we don't need to go into greater details, in the present chapter, to conclude, as many distinguished biologists do these days, that even if mutations were really random at their source, the corresponding phenotypes are not. In other words, before any phenotype can be, so to speak, 'offered' to selection by the environment, a host of internal constraints have to be satisfied and, as we are going to see, interactions at many levels have to be stabilized. A variety of filters, some acting in series, some cooperating or interfering, stand between mutations and their expression.

There is, in short, no single 'arrow' connecting a random generator of genetic diversity to the phenotypes on which exogenous selection acts. There are different effects of different kinds of filters and regulatory processes, at different levels, presently under intense scrutiny (Mattick, 2005; Amaral and Mattick, 2008; Mattick and Mehler, 2008). There usually are differential rates of efficiency for the different variants, at each level, and different kinds of local (that is, endogenous) selections. There are also exogenous selections, but here too the story is quite different from the one offered by standard neo-Darwinism, as we will see in a moment. Some evolutionary biologists have, in fact, generalized and expanded the mechanisms of Darwinian selection to include internal selection.[31] Part two will explain the conceptual shortcomings of Darwinism that also apply to these Darwinian expansions, but before we go into that, several other facts and new developments in biology proper have to be taken into account. The picture of the relation between genes and phenotypes becomes even more complicated when we look at the next family of levels up: that is, the relations between the genome as a whole and pathways of development. We turn to these in the next chapter.

3

WHOLE GENOMES, NETWORKS, MODULES AND OTHER COMPLEXITIES

Gene regulatory networks

Extremely complex gene regulatory networks are at work in the developing organism and they offer important new keys to the origins of animal body plans and evolution (Davidson, 2006; Davidson and Erwin, 2006; de Leon and Davidson, 2009).[1] Davidson and Erwin (2006) argued that known microevolutionary processes cannot explain the evolution of large differences in development that characterize entire classes of animals.[2] Instead, they proposed that the large distinct categories called phyla arise from novel evolutionary processes involving large-effect mutations acting on conserved core pathways of development. Gene regulatory networks are also modular in organization (Oliveri and Davidson, 2007).[3] This means, in essence, that they form compact units of interaction relatively separate from other similar, but distinct, units.

The consequence is that these processes make the connection between specific biological traits, specific evolutionary dynamics and natural selection very complicated at best, impossible at worst. In the words of a leading expert of gene regulatory networks:

Developmental gene regulatory networks are inhomogeneous in structure and discontinuous and modular in organization, and so changes in them will have inhomogeneous and discontinuous effects in evolutionary terms ... These kinds of changes imperfectly reflect the Class, Order and Family level of diversification of animals. The basic stability of phylum-level morphological characters since the advent of bilaterian assemblages may be due to the extreme conservation of network kernels. *The most important consequence is that contrary to classical evolution theory, the processes that drive the small changes observed as species diverge cannot be taken as models for the evolution of the body plans of animals.* These are as apples and oranges, so to speak, and that is why it is necessary to apply new principles that derive from the structure/function relations of gene regulatory networks to approach the mechanisms of body plan evolution.

Davidson, 2006, p. 195, emphasis added

Additional phenomena, such as developmental modules, entrenchment and robustness, further separate random mutations at the DNA level from expressed phenotypes at the level of organisms. We will develop the idea of developmental and evolutionary modularity in a moment. Let's first briefly characterize entrenchment and robustness.

Entrenchment

The different components of a genome and/or of a developmental structure usually have different effects 'downstream', that is, on the characteristics of the fully developed adult, through the entire lifetime. The magnitude of these effects is measured by the 'entrenchment' of that structure. The entrenchment of a gene or a gene complex changes by degrees – it's not an all-or-none property. From an evolutionary point of view, the entrenchment of a unit has multiple and deep consequences for its role in different groups of organisms and different species, notably affecting other units that depend on its functioning. Generative entrenchment (Wimsatt, 1987; Schank and

Wimsatt, 2001; Wimsatt, 2003) is seen both as an 'engine' of development and evolutionary change, and as a constraint. This amounts to saying that crucial developmental factors ('pivots' in Wimsatt's terms) may be highly conserved and be buffered against change, or may undergo minor heritable changes with major evolutionary consequences. Generative entrenchment, as the expression aptly suggests, is very probably linked to spontaneous and quite general collective form-generating processes that we will review in the next chapter, but it is (of course) also under the control of genes, gene complexes and developmental pathways. How these different sources of order and change (some generically physico-chemical and some specifically genetic) interact is still largely unknown (Kauffman, 1987, 1993).

Robustness

A trait is said to be robust with respect to a genetic or environmental variable if variation of the one is only weakly correlated with variations in the other. In other words, robustness is the persistence of a trait of an organism under perturbations, be they random developmental noise, environmental change or genetic change. Many different features of an organism, both microscopic and macroscopic, could qualify as traits in this definition of robustness. A trait could be the proper fold or activity of a protein, a gene expression pattern produced by a regulatory gene network, the regular progression of a cell division cycle, the communication of a molecular signal from cell surface to nucleus or a cell interaction necessary for embryogenesis or the proper formation of a viable organism or organ, for example (Felix and Wagner, 2008). Robustness is important in ensuring the stability of phenotypic traits that are constantly exposed to genetic and non-genetic variation. In recent years, robustness has been shown to be of paramount importance in understanding evolution, because robustness permits hidden genetic variation to accumulate. Such hidden variation may serve as a source of new adaptations and evolutionary innovations (Kitano, 2004).

The source of robustness lies in the fact that the developmental processes that give rise to complex traits are nonlinear (Nijhout, 2002). In a recent paper, two leading experts say:

> A consequence of this nonlinearity is that not all genes are equally correlated with the trait whose ontogeny they control. Because robustness is not controlled independently from the core components of a system, *it is not straightforward to disentangle buffering mechanisms that have been subject to natural selection from those that have not*. This is a major challenge for future work.
>
> Felix and Wagner, 2008, emphasis added

In 2005, reviewing a book on robustness and evolvability by Andreas Wagner (Wagner, 2005) in *Science*, Gregory C. Gibson, the William Neal Reynolds Distinguished Professor of Genetics at North Carolina State University, says:

> Robustness must involve non-additive genetic interactions, but *quantitative geneticists have for the better part of a century generally accepted that it is only the additive component of genetic variation that responds to selection*. Consequently, we are faced with the observation that biological systems are pervasively robust but find it hard to explain exactly how they evolve to be that way.
>
> Gibson, 2005, p. 237, emphasis added

In what is music to our ears, in this review Gibson adds, '[this book] contributes significantly to the emerging view that natural selection is just one, and maybe not even the most fundamental, source of biological order.'

Darwin himself had explicitly acknowledged that natural selection is not the only mechanism in evolution, but it's worth stressing that these days, as Gibson prudently (with 'maybe') says, it's 'not even the most fundamental one'. We want to go further along this path and conclude that these multiple levels of internal constraints on possible phenotypes make the notion of evolution as the product of external selection operating on phenotypic variations generated at random radically untenable.[4] Darwin argued that (to borrow Dennett's phrase) phenotypes 'carry information about' the ecologies in which they evolved. The brown colour of the butterfly tells us that it evolved in a smoky atmosphere.[5] But it now seems undeniable that evolved phenotypes also carry information about the *internal* organization of the creatures that have them (about their genotypic and ontogenetic structures, for example.) It is an open, empirical and highly substantive question how narrowly such endogenous effects constrain the phenotypic variations on which external selection operates. It will take a while to find out. But, until that question gets answered, it is unadvisable to take a neo-Darwinist account of evolution for granted.

Master genes are 'masters'

Many different traits are indissociably genetically controlled by the same 'master gene' (this is technically called pleiotropism – from the ancient Greek, meaning 'motion in many directions'). Any mutation affecting one master gene, if viable, has an impact on many traits at once. Moreover, new variants of a trait may interact differently with variants of other traits. The timing and intensity of expression of genes are, as we saw, controlled through complex gene regulatory networks (Coyne, 2006; Davidson and Erwin, 2006; Erwin and Davidson, 2006). An important consequence of genetic pleiotropism is that, when a gene affects several traits at once, any change in that gene that is not catastrophic (any viable mutation) will affect all or most of these traits. Supposing that one such change in one such trait is adaptive, then natural selection will eventually fixate that mutation. But then all the other changes in all the other traits will also be stabilized, possibly opening up wholly different selective processes, eventually dwarfing the effects of the initial selection driven by the initially adaptive trait.[6]

There is an interesting example that we choose here, tentative as it may be, because it concerns the evolution of brain and therefore of cognition. It has been suggested that there are regulatory genes that affect many different organs, including the development of the cerebral cortex (Simeone, 1998; Simeone *et al.*, 1992, 1993). A well-studied gene family, called Otx, masterminds the development of kidneys, cranio-facial structures (Suda *et al.*, 2009), guts, gonads and the cerebral cortex (segmentation and cortical organization). Several mutants are known, including severe pathological cases in humans (at one extreme lissencephaly – an abnormally smooth brain surface – at the other schizoencephaly – an exaggeratedly deep inter-hemispheric cleft). Mutants are usually short lived and leave no progeny.

Italian geneticist Edoardo Boncinelli (Boncinelli, 1998, 2000) has offered an interesting and relatively tentative hypothesis which, if even roughly correct, implies that there are significant aspects of our brain structure that are not consequences of selection for their fitness but rather side effects of selection for quite other phenotypic traits (spandrels in Gould and Lewontin's sense; see Chapter 6); in particular, since the Otx1 'master' gene controls the development of the larynx, inner ear, kidneys and external genitalia and the thickness of the cerebral cortex, selective pressures sensitive to changes in the functions of the kidneys (due to the bipedal station, or different liquid intake and excretion resulting from floods or droughts), or the fixation of different sexual patterns, may have had in turn secondary effects on the expansion of the cerebral cortex and the structure and function of the larynx. The peculiarity of the overall picture of the evolution of language and cognition in humans, should this reconstruction prove to be correct, has been stressed to us by Boncinelli (personal communication, June 2009). Neither we nor Boncinelli are claiming that this actually is *the* right evolutionary story about the emergence of the enlarged cortex in the human brain, only that some such story *might* be correct and that it is, as far as we know, consonant with the facts currently available. A dogmatic adherence to adaptationism blinds one to such interesting possibilities.

Moreover, it's known today that, just as the same phenotype may be the result of quite different genes or gene complexes (convergence), different phenotypes may be the result of the same genes or gene complexes (differential gene regulation). The epigenetic effects that we mentioned earlier, and to which we will return shortly, may mimic genetic ones, with drastically different consequences for the degree of plasticity of that trait, and/or different possibility of its fixation and susceptibility to further variation and evolution. The effect of the genetic context as a whole upon a new variant may be one of suppression (negative epistasis), of enhancement (positive epistasis) or compensation (compensatory epistasis), with quite different effects on its contribution to overall fitness (Pigliucci, 2009b). Moreover, a general non-lethal decrease of average fitness in a population may 'turn' a disadvantageous mutation into an advantageous one (Silander et al., 2007).[7]

Developmental modules

Let's start with a definition.[8] A module is a unit that is highly integrated internally and relatively insensitive to context externally. Developmental modules exist at different levels of organization, from gene regulation to networks of interacting genes to organ primordia. They are relatively insensitive to the surrounding context and can thus behave invariantly, even when they are multiply realized in different tissues and in different developmental phases. Different combinations of developmental modules in each context, however, produce a difference in their functions in development. There is evidence of the integration of several interacting elements into a module when perturbation of one element results in perturbations of the other elements in that module, or in gene–gene interaction (epistasis) within the module, in such a way that the overall developmental input–output relation is altered. This is another signal case in which the conservation of genetic and developmental building blocks, together with their multiple recombinations in different tissues and organisms, explains the diversity of life forms as well as the invariance of basic body plans.[9] The double-edged (so to speak) character of developmental modules consists in their relative context insensitivity to external factors and their relative context sensitivity to some internal substitutions of subcomponents.[10] This is presently a very active and very complex domain of inquiry (for a vast survey see the volume edited by Schlosser and Wagner, 2004). In evolution, developmental modules may preserve their integrity in spite of being embedded into different heritable variations of their context and also, in several cases, in spite of the replacement of some of their sub-modules by others. Gerhard Schlosser writes: '[Developmental modules] may form coherent and quasi-autonomous units in evolution (modules of evolution) that are repeatedly recombinable with other such units' (Schlosser, 2004, p. 520).

In essence, the 'logical' role of a module is one of presenting cascades of interacting elements, where the output of one provides some of the input to the others. Developmental modules are triggered in a switch-like fashion by a variety of inputs, to which they are only weakly linked. This weak linkage admits variations and allows relatively novel inputs. These inputs are 'triggers' (*sic*, in this literature) not templates of shapes. The way in which modules affect different downstream processes depends on the overall genetic context. It's worth stressing that the internal machinery is predisposed to react in complex ways to a class of switches. All this makes the development of organisms an intricate network of context-independent processes (the modules) and of internally context-dependent ones (interactions between modules and interactions of the modules with other structures). The reverberation of the effects of gene mutations is usually multiple and only the viable overall result is then accessible to selection.

There are different classes of modules. The most basic and earliest operating class affects the regulation of gene transcription at distinct but interacting levels. As a consequence, DNA sequences that act as promoters and enhancers can be swapped between genes. It is also the case that multiple enhancers exist for a single gene, each controlling a particular expression domain of the gene. These can be multiply recombined. The basic transcriptional apparatus (BTA) is itself modular, and its specificity can be changed by swapping different transcription factors.

An especially interesting class of modules are the signalling pathways, families (or classes) of proteins acting in concert in cascades that constitute whole cycles or networks, and representing biochemical 'signals' that have specific types of cells as their targets in different issues, such target cells often lying side by side with unresponsive (non-target) cells. Only five major families appear to be important during early embryonic development (because of their [separate] initial discoveries, they bear names that sound bizarre to the uninitiated: hedgehog, TGF, Wnt, receptor tyrosine kinases [RTKs] and Notch). Each family is relatively autonomous with respect to the others; each class has its own primary role, but many can also play multiple roles in the development of very different tissues. For example, the Notch system also acts as a positive feedback loop between neighbouring cells, amplifying initial differences (determining different fates of neighbouring cells). This complex system of master signals regulates tissues as different as the central nervous system, pharynx, hair cells, odontoblasts, kidney, feathers, gut, lung, pancreas, hair and ciliated epidermal cells across many different vertebrate and invertebrate species. Every mutation in any one of the genes involved will alter many organs and their functions – a far cry from 'beanbag genetics'.

Organ primordia such as limb buds and mandible and teeth primordia act like modules (Zelditch *et al.*, 2008) and can be transplanted to develop ectopically – that is, in different, non-canonical parts of the embryo.[11] This can happen partially, to a certain extent, or even completely, in diverse parts of the embryo, with different results in different species. As a rule of thumb, the transplantation and activation of genes across species, or out of place (ectopically) in the same species, is more successful for genes that are normally expressed sooner in the life of the embryo than for genes that are expressed later. This, as Stephen Jay Gould and Brian Goodwin have argued, gives some, only some, substance to the old idea (originally due to K. E. von Baer and Ernst Haeckel) that ontogenesis recapitulates phylogenesis (the successive forms of the developing embryo are reminiscent of the ascent of forms in evolutionary time).

Some modules are systemic modules, distributed throughout the organism. The best examples are hormonally mediated processes, in which only a subset of cells in various tissues is responsive to a particular hormone, intermingled with unresponsive cells. Nonetheless, the response is substantially the same everywhere, with many orchestrated changes: new metabolic enzyme expressions are switched on; likewise extensive programmed cell death (or its inhibition), and the differentiation of new cell types (gut, epidermis); likewise the remodelling of muscles, and of parts of the nervous system. The thyroid-hormone-dependent metamorphosis in amphibians is modulable at will by mere administration of various doses of the hormone.

The lesson here is that modularity gives a new complex picture of evolution, one in which internal constraints and internal dynamics filter what selection can act upon, and to what extent it can do so. Precisely because so much cannot change, other things can change at the (so to speak) genetic periphery of organisms. It is often (although not always) the case that when we witness gene duplications, a ubiquitous kind of genetic modification, the 'original' gene continues acting as it did in earlier forms of life, while the 'copy' can 'explore' new functions over evolutionary time (these metaphors are commonplace in the professional literature).

Coordination

We saw earlier how badly misguided the additive, 'beanbag' conception of genes is. There is more to be said about this. The Russian zoologist and evolutionist Ivan Ivanovich Schmalhausen (1884–1963) had rightly stressed that living organisms are not the mere atomic 'adposition' of separate parts, but rather highly 'coordinated' systems (for a historical and critical review, see Levit *et al.*, 2006). Today justice is done to Schmalhausen by experimental evidence that some mutations in genes specifically affecting one part of the body carry with them suitable modifications in other related parts. When limbs are induced ectopically (that is, where they don't belong), often sensory neurons, receptor organs, cartilage and blood vessels also develop as a consequence around them (see Kirschner and Gerhart, 2005 for stunning examples). A laboratory-induced and quantitatively controllable modification in two key proteins[12] in chick and finch embryos early in development produces as the main result variable elongation and thinning of the upper part of the beak (Abzhanov *et al.*, 2006). However, the lower beak and the neck muscles also 'follow'.

The lesson here is, once again, that natural selection cannot select isolated traits, but rather coordinated complexes of traits, coming all together in virtue of pleiotropism, developmental solidarity (Schmalhausen's coordination) and epigenetic modifications (see below).

Morphogenetic explosions

After what we have seen in this rapid and summary exposition, it stands to reason that, in consequence of the many internal constraints on possible new life forms, when one or more of these constraints are internally, genetically, relaxed or withheld, new possibilities open up, sometimes in an explosive way (Gould, 1989, vindicated in Erwin, 2008 and Theissen, 2009). Over periods that are relatively short in geological terms, a great variety of new life forms appears suddenly and (as palaeontologists say) 'explosively'. This seems to have happened at least twice in the remote past, and at least once more recently.

The Ediacara fossils (from 575 to 542 million years ago) represent Earth's oldest known complex macroscopic life forms. A comprehensive quantitative analysis of these fossils indicates that the oldest Ediacara assemblage, the Avalon assemblage, already encompassed the full range of the possible forms of the Ediacara (what is technically called their 'morphospace', the repertoire of accessible forms) (Erwin, 2008; Shen *et al.*, 2008). A comparable morphospace range was occupied by the subsequent White Sea assemblage (560 to 550 million years ago) and Nama assemblage (550 to 542 million years ago), although it was populated differently (taxonomic richness increased in the White Sea assemblage but declined in the Nama assemblage). These changes in diversity, occurring while the range of forms (the morphospace) remained relatively constant, led to inverse shifts in morphological variance. The Avalon morphospace expansion may well mirror the Cambrian explosion (about 545 million years ago) when in the relatively short period of 5 to 10 million years most of the complex life forms we see today appeared on Earth, and both events may reflect similar underlying mechanisms. The palaeontologist Douglas H. Erwin (2008) summarizes these findings, saying that they amount to the recognition that these ancestral forms of life apparently contained already a suite of developmental tools for differentiating their body plans, although not yet the sophisticated developmental tools capable of building the regional body patterning of higher animals. For that, we will have to wait until the momentous Cambrian explosion.

Another example of morphological explosion and of its importance in the unravelling of poorly understood macroevolutionary processes is analysed in Moyle *et al.* (2009). The relatively recent (between two million and a million and a half years ago) explosive Pleistocene diversification and hemispheric expansion of 'white-eye' passerine birds (Zosteropidae, a family containing among the most species-rich bird genera) represents a per-lineage diversification rate among the highest reported for vertebrates (estimated to be between 1.9 and 2.6 species per million years). However, these authors stress that, unlike the much earlier explosions seen above, this rapid rich diversification was not limited in geographic scope, but instead spanned the entire Old World tropics, parts of temperate Asia and numerous Atlantic, Pacific and Indian Ocean archipelagos.

Interestingly, this paper reports that the tempo and geographic breadth of this rapid radiation 'defy any single diversification paradigm, but implicate a prominent role for lineage-specific life-history traits (such as rapid evolutionary shifts in dispersal ability) that enabled white-eyes to respond rapidly and persistently to the geographic drivers of diversification' (Moyle *et al.*, 2009).

By means of a comparative analysis of sequences of nuclear and mitochondrial DNA, a small group of ancestors characterized as 'great speciators' (Diamond *et al.*, 1976) has been extrapolated. This 'hyperdiversification' can only be explained via a complex interaction between intrinsic and extrinsic drivers of rapid speciation, combining with processes of reproductive isolation and migration. These authors conclude that 'the pattern and tempo of diversification recovered for the white-eyes do not fit comfortably within any single diversification paradigm (e.g., dispersal, vicariance, equilibrium island biogeography, etc.) and underscore the importance of casting a broad net, in terms of taxonomy, geography, and theory, in modern diversification studies' (Diamond *et al.*, 1976).

We can summarize by saying that morphological explosions may well reflect major changes in internal constraints as crucial components in speciation. If so, then the effects of natural selection may well consist largely of post-hoc fine-tuning in the distribution of subspecies and variants (Newman and Bhat, 2008): quite a different kind of account[13] from the one of gradual selection of randomly differing small variations.[14]

Plasticity and the (non)transitivity of fitness

This survey of stumbling blocks for gradualism (that is, for the thesis that many intermediate variants must have existed between any two significantly different evolutionarily related forms) is by no means complete.[15] We now propose to add a few more. The first is that Darwinian fitness is heavily dependent on the context in which it is assessed. The second is that, as a consequence, relative fitness does not transfer from one comparison to another, even within the same species. The third is that not only phenotypes but also genomes can be plastic. The fourth and, for the moment, final is that the textbook cases of Mendelian inheritance, in spite of their great historical and didactic importance, are more the exception than the rule.

A crucial, though usually tacit, assumption of gradualism is that adaptive modification is transitive. If variant A has greater fitness than variant B, and B has greater fitness than C, then A must have greater fitness than C. There is no gradualist adaptive story to be told unless the process is assumed to be transitive.[16] This leads to a perennial puzzle about how evolution avoids being trapped in local maxima of fitness.[17] In fact, gradualist-selectivist adaptationism characteristically depicts the evolutionary process as one of hill climbing, not infrequently deploying, as we said earlier, pleasant graphic artistry to convey the intuitions behind the model of fitness landscapes and adaptive landscapes. Although the possibility that an organism (or a population) has been 'trapped' on top of a local maximum, in the proximity of a nearby, but inaccessible, higher peak has long been acknowledged in classical neo-Darwinism, with few exceptions the climbing process itself has always been assumed to be smooth, and each path locally transitive. However, where there have been morpho-genetic explosions, fitness relations become surely non-transitive and plausibly irrelevant. There is no hill climbing, not even a smooth path from each level of fitness to the next; only a jumpy traverse of a maze[18] or a 'glass-like' surface with a huge number of neighbouring peaks.

In fact, numerous instances of non-transitive differentials of fitness have been observed. For instance, when one variant A only competes with another variant B, then A leaves more descendants and 'outsmarts' variant B. The same occurs when variant B only competes with variant C. But if all three compete, then it may well be that C wins over A.

The relation between genotypic variation and phenotypic plasticity suggests that we could consider genes as 'norms of reaction' to different environments (Lewontin et al., 2001). Plotting the measure of a trait against environmental parameters usually gives complex curves, complex 'norms of reaction' that do not reveal transitive relations of fitness. For instance, out of about a dozen variants of the plant Achillea, one is the tallest at low altitudes and at high altitudes, but the shortest at intermediate ones (Suzuki et al., 1981). Mapping the size of such variants of Achillea over heights ranging from sea level to 3,000 meters, we obtain quite 'bumpy' curves that criss-cross one another. The same intricate zigzagging geometry also applies to the curves plotting total leaf area, total leaf number and mean leaf size as a function of soil moisture in the plant Polygonum persicaria (spotted ladysthumb) (Sultan and Bazzaz, 1993). In the animal kingdom, similar criss-crossing curves for different genotypes are to be found, for instance in the number of abdominal bristles in the fruit fly, as a function of the ambient temperature during development, and in the rate of production of specific membrane proteins in the immune system for different human genotypes as a function of early exposure to domestic animals (Eder et al., 2004; Martinez, 2007; von Mutius, 2007).

The study of these plastic phenotypes and 'plastic genomes' [*sic*][19] has been described as an 'alternative picture' to a conventional view of precise genotype-specific adaptations (Bradshaw and Hardwick, 1989).[20] It was becoming apparent that the textbook cases of natural selection (such as the much-debated and rather dubious case of industrial melanism and the peppered moths, *Biston betularia*) do not exemplify the ordinary evolutionary processes in ordinary environments (Pigliucci, 2009a), especially for complex traits that are under the governance of many genes. The departure from textbook cases of natural selection and of straightforward Mendelian inheritance (one single gene mutation producing one pathology) have been superseded in recent years by many-faceted gene–environment interactions for a variety of complex traits and complex diseases (for instance in the case of asthma) through multifactorial analyses, with statistical methods and biological mechanisms yet to be exhaustively characterized (Vercelli, 2008). The field is in constant motion, but there is consensus that the evolutionary role of plastic traits and plastic genotypes departs from the standard neo-Darwinian gradualist-adaptationist picture in many ways. Continuous hill climbing and transitive fitness relations are, at best, more the exception than the rule.

Conclusions and a caveat

So much for a brief survey of the trees – we very much do not want it to obscure the view of the forest. The point to keep your eye on is this: it is possible to imagine serious alternatives to the traditional Darwinian consensus that evolution is primarily a gradualistic process in which small phenotypic changes generated at random are then filtered by environmental constraints. This view is seriously defective if, as we suppose, the putative random variations are in fact highly constrained by the internal structures of the evolving organisms. Perhaps it goes without saying that the more the internalist story is true, the less work is left for appeals to natural selection to do. In fact, as Ron Amundson (2006) has aptly remarked, neo-Darwinists have an unfortunate tendency to view internal constraints as 'idealistic'. For reasons that we will make clear in Part two (especially Chapter 6), in psychology, behaviourists manifested a similar reaction, opposing all appeals to internal cognitive structures and constraints. If that's 'idealism', well, so be it. We rather think it's just ordinary naturalism.

A caveat. It's only fair to acknowledge that the majority of biologists whom we have cited here, including several of the discoverers of these quite intricate levels of endogenous regulation, still today endorse natural selection[21] as the determinant par excellence of the course of evolution. Indeed the most determined defenders of neo-Darwinism consider the sorts of results we've been surveying as further supporting natural selection. (We confess to not understanding that, but see the crisp exchange about evolutionary psychology between, on the one side, Robert Lickliter and Hunter Honeycutt (2003a, 2003b) and, on the other, John Tooby, Leda Cosmides and Clark H. Barrett (Tooby *et al.*, 2003).[22] Lickliter and Honeycutt are accused of 'present[ing] routine findings and viewpoints that have been generally accepted for decades as if they constituted a 'conceptual revolution' that has 'transformed contemporary developmental and evolutionary theory.' We expect to be accused of that too. But bear in mind: simply to acknowledge results that are counter-examples to one's theory is not to make the results go away. One thing that happens to theories that hang around past their time is that they're nibbled to death by 'routine findings'. Moreover, we have frequently noticed, both in the literature and in conversation, a lurking confusion between what is compatible with, or supports, evolution and what is compatible with, or supports, neo-Darwinian adaptationism (for a clear explicit case, see Coyne, 2009). Adaptationism is the (putative) mechanism of evolution. It is entirely compatible with what we've been saying that Darwin was right about the one but wrong about the other. That, indeed, is what we're betting on.

However, at this stage of our presentation, we are prepared to settle for a stand-off. The sorts of results we've been reviewing suggest that the case for neo-Darwinism is less than apodictic. Adaptationism isn't what they call in the Midwest 'just a theory', but it isn't a dogma either. So be it; more arguments are needed, and we have more up our sleeves. We'll argue presently that, quite aside from the problems it has accommodating the empirical findings, the theory of natural selection is internally flawed; it's not just that the data are equivocal, it's that there's a crack in the foundations.

4

MANY CONSTRAINTS,
MANY ENVIRONMENTS

In recent years, several pleas have been made for considering selection at many levels, and for revising the traditional neo-Darwinian thinking that either the individual, or the population as a whole, are the sole units of selection. There are selective processes and competitions and synergies also at the level of genes, chromosomes, whole genomes, whole epigenomes, cells, developing tissues, kin groups, societies and communities; and, of course, organisms and populations. Each one of these levels shows specific dynamics and interface phenomena with the levels immediately below and above (Michod, 1999).

This chapter summarizes some of them.

Selection without adaptation

Along with the recent discoveries and developments that we have just reviewed, we also wish to mention a conundrum (for the canonical neo-Darwinians) that has been on the scene for decades, although mostly neglected until very recently. The interesting saga leading to the discovery of the chaperone HSP90 mentioned earlier has equally interesting antecedents that deserve to be told, at least in essence.

The aforementioned British geneticist and embryologist Conrad Hal Waddington showed in the 1950s that, under unusual environmental conditions, by repeated artificial selection, new phenotypes can emerge that have no evident adaptive relation to these environments. For instance, a small fraction of wild-type strains of fruit flies repeatedly grown in the egg stage in the presence of fumes of ether develop (an initially quite small) second pair of wings. By repeatedly selectively cross-breeding these variants, one generation after the other, and repeatedly submitting the progeny to fumes of ether in the egg stage, a progressively greater number of individuals can be produced that develop this additional pair of wings (resulting in two pairs of wings, instead of one). The second pair consists of small winglets, rather than the tiny wing buds of the wild type, called halteres.[1] Eventually, after many generations so treated and cross-bred, Waddington noticed that individuals with the aberrant additional pair of wings could produce offspring that no longer needed ether egg shock to produce the wings. This new phenotype had become stabilized. Let's notice that it would have been quite impossible for natural selection to create this second pair of wings, because there is no spontaneous variation in this character. Similar procedures were applied, and similar results were obtained, by Waddington and others on a variety of different traits. For instance, another interesting variant appears when wild-type fruit flies are raised at high temperature and are again cross-bred for generations. This other variant is characterized by unusually large eyes (Waddington, 1956, 1957). After selective crossbreeding over several generations, these phenotypes are stabilized too and are observed even when the offspring of the selected fruit flies are raised in a normal environment. Waddington called this phenomenon 'genetic assimilation'.

A general characterization of this story is the following[2]: a character has no variation normally, but if you shock the developmental system in suitable ways, then a few individuals with an abnormal trait appear. If you breed from those individuals, and again shock their offspring early in development, the effect becomes more pronounced and more and more of them are affected, one generation after the other. You eventually get enough modified individuals and can breed for the abnormal character without further shocking.[3] There is no adaptive

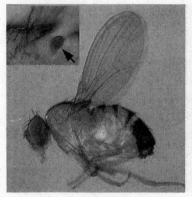

Interchangeability between reactions to external and internal perturbations. The extra small pair of wings shown here (right) can be either the result of an (internal) mutation in a master gene (called Ubx, for ultra-bi-thorax) or of Waddington's (external) ether treatment described in the text. © Nipam Patel, from Barton et al.

process in sight, and the relation between these environments and these phenotypes is, to put it mildly, elusive.[4]

These data have only recently turned out to be most revealing from a genetic, epigenetic and developmental point of view (Rutherford and Lindquist, 1998; Queltsch *et al.*, 2002; True *et al.*, 2004; Sangster *et al.*, 2008). Waddington had unearthed the presence of silent mutations accumulating in what are now called 'evolutionary capacitors'. In spite of some differences in the experimental procedures, the outcomes are very similar. In fact, Waddington applied an unusual growth milieu (ether or high temperature) and then selective breeding for many generations, while Lindquist and colleagues mutated the gene for HSP90 pharmacologically. In both cases, the expression of the related phenotypes is today of central importance to genetics, and to epigenetics. What we wish to stress is the lack of any transparent adaptive correlation between such phenotypes and the selecting environment. Terms such as 'canalization' (due to Waddington) and 'capacitor' (due to Lindquist, Sanger and Rutherford) reveal a considerable change with respect to the canonical neo-Darwinian picture.

The achievement of neo-Darwinian explanations is supposed to consist precisely in starting with a story of blind trials and errors, and deriving, at the end of the day, a 'brilliant', 'exquisite' match between the evolved phenotype and the requirements of the external environment.[5] By contrast, the sorts of cases reported here do not suggest an adaptationist treatment; in consequence, they were largely forgotten until the emergence of epigenetics revived an interest in Waddington's experiments. The lesson from such cases is that unusual environments and selective processes in them (artificial in this case, natural in others) do produce various phenotypes, of which some are not lethal, but not adaptive either. It bears emphasis that the 'information' that such phenotypes give about the evolutionary environment in which they emerged is very, very indirect, or nil.

Constraints on selectability and the rise of contingency

There are, as we have seen, all sorts of internal constraints on adaptation. There are also external constraints that do not operate by natural selection. This has long been acknowledged in population genetics, wherein standard textbooks emphasize the fixation of phenotypic traits in small populations by random genetic drift, frequency-dependent selection, density-dependent selection and so forth.[6] There are, for example, well-examined, clear cases of when fitness is a function of (among many other variables) overall population size, and when there is 'the ascent of the abundant' – that is, when abundant phenotypes may acquire an evolutionary advantage regardless of their fitness (Cowperthwaite *et al.*, 2008). The very complex interplay of all these factors, of internal constraints, internal selection and external selection, is arguably the very core of evolution which is a different and vastly more complex story than the one told by classical neo-Darwinism.

Bacteria, sugars and dead ends

It is worth the effort to consider a paradigmatically simple case, discovered serendipitously, in the most famous of all experimental biological models: the intestinal bacterium *Escherichia coli*. It is paradigmatic because a bacterium is the simplest self-reproducing organism there is (viruses need bacteria or cells to reproduce), because the processes at work are so elementary, and because the function on which selection converged is the most straightforward one can imagine: the enzymatic digestion (fermentation) of sugars (Hall, 1978, 1981).[7]

The story is as follows: in the naturally occurring variant (wild type) of *E. coli* we find an enzyme (called ebgA) that 'digests' (ferments) the naturally occurring sugar lactose (let's call it sugar 1) but cannot digest similar, but distinct, sugars (lactulose – let's call it sugar 2 – and lactobionate – let's call it sugar 3). There is also a gene (*ebgR*) that regulates the synthesis of this enzyme. The initial aim of a direct (let's notice, direct) laboratory selection was to obtain a variant of the bacterium that could also ferment sugar 3. But this failed. A backtracking procedure was then followed. The attempt was made to select, one step at a time, bacterial strains that would ferment sugar 2. Various quite ingenious methods of serial selection were tried out, and the desired result was eventually obtained, but many 'dead ends' [*sic*] were also produced. In particular, after having selected strains that carry the desired (unregulated) variant of the regulatory gene (obtaining strains that could digest both sugar 1 and sugar 2), attempts were made to directly select strains that could digest sugar 3 as well. But no luck; the glitch is that the strains capable of fermenting both sugar 1 and sugar 2 never give mutants that can also digest sugar 3. In fact, if one selects from a sugar-1 strain, one must screen away and dump all mutants that can also utilize sugar 2, otherwise one will never obtain any mutants that can utilize sugar 3. Notice that these bacteria have never encountered sugar 2.[8]

In effect the sugar-1 and -2 strains are dead ends towards the utilization of sugar 3. The order of the selective steps cannot be altered. Moreover, it turns out that some optimally effective mutants for the old function are dead ends for further adaptation to the new function. That is, selection for a new function (by a planned change of the growth medium) sometimes reduces the efficiency of the old function. It is not the case that the best adapted to the old function is also the best adapted to the new one, nor is it the case that the converse applies – that is, that the best adapted to the new function is also the best adapted to the old one.

Some lessons from this paradigmatic ultra-simple case

In the story above, we know exactly what was being selected, how, when and 'for' what, because it was an artificial selection made by smart experimentalists and published in detail. Yet, other (undesired, in this case) traits were also being selected, with no possibility of doing otherwise. Selection with just sugar 1 gives as a result some strains of bacteria that also carry the capacity to digest sugar 2. In a well-established terminology, to which we will return in what follows, this second capacity is a 'free-rider' on the first. Had it not been for the unanticipated block that it represents towards the additional utilization of sugar 3, it would have been a welcome new trait. On paper, this additional capacity is advantageous; being able to use two types of sugar instead of only one looks like an evolutionary plus. Too bad, though, that it precludes all possibility of also using sugar 3. In the laboratory, these 'free-riders' were detected and discarded; in a laboratory environment, coextensive phenotypic traits can be experimentally dissociated to determine which of them is actually adaptive and which merely free-rides. But, of course, natural selection doesn't perform experiments; if phenotypic traits are coextensive in the environment of selection, then, if either is selected, both are.

Let's repeat, for the sake of clarity, that those bacteria had never been exposed to sugar 2, and free-riding would have been undetectable in the initial selective medium (sugar 1). We will return to this point (see Chapter 6). Suffice for now that there is an interesting and transparent analogy with free-riding phenotypes in the case of domestication. Many traits that were never directly selected for by breeders appear in domesticated species. These traits 'free-ride' on domestication but, unlike the 'dead ends' in the case of *E. coli*, had no reason to be discarded. Domesticated species as different as sheep, poodles, donkeys, horses, pigs, goats, mice and guinea pigs, each one of them artificially selected for some specific trait (quality of wool, friendliness to humans, loadability in transportation, and so on), also present dwarf and giant varieties, piebald coat colour, wavy or curly hair, floppy ears and more (Trut, 1999).

Out in nature things are never as simple as they are in a laboratory, or on a farm; natural selection is whatever it happens to be, and it's hard, often impossible, to reconstruct what trait (if any) has been naturally selected, and 'for' what, as distinct from traits that merely free-ride (see Chapter 6). Past natural situations capable of 'splitting' coextensive phenotypes may have never happened, or may have happened leaving no trace. Dead ends are prone to be just that: dead. And we will never know whether they existed, what they looked like if they did exist, or why they were doomed. Counterfactual reasoning (what would have been the case if ...) is unavailable in cases in which controlled experiments cannot be run. Moreover, the order of the selection steps may well be as crucial in nature as it was in this experiment, and the final result may well be strictly dependent on past contingencies.[9] In this case, we are told which steps were taken in which order. Lucky us! But again, out in nature, there is usually no way for us to reconstruct the steps and their order. The moral here is that the application of adaptationism even to specific elementary cases such as a bacterium and the digestion of three sugars presents formidable problems. So far, these cases are not telling us that natural selection did not happen, but they tell us that the parallel between artificial selection and natural selection, so central to Darwin's theory, is flawed (a point to which we will return in Part two). They also tell us that our capacity to reconstruct the vagaries of natural selection is extremely limited, to say the least. The vastly more complex case of the evolution of higher multicellular species, with all the intricate internal constraints and gene interactions, often makes reconstruction impossible unless model systems (actual species) can be genetically manipulated in a laboratory and strong analogies to plausible evolutionary events can be drawn (for one of the earliest and most persuasive such reconstructions, see Ronshaugen et al., 2002).

New kinds of environments

In a centenary symposium in honour of Ernst Mayr (but with a quite different perspective from Mayr's), Mary-Jane West-Eberhard, a prominent representative of 'evo-devo', focuses on developmental plasticity and invites us to consider a quite different kind of 'environmentalism'; not the traditional one in which an all-powerful selective filter winnows pre-existing genetic variability, but an environment that initiates genetic changes by amplifying developmental plasticity (a topic to which we will return):

> Change in trait frequency involves genetic accommodation of the threshold or liability for expression of a novel trait, a process that follows rather than directs phenotypic change. Contrary to common belief, environmentally initiated novelties may have greater evolutionary potential than mutationally induced ones. Thus, genes are probably more often followers than leaders in evolutionary change. Species differences can originate before reproductive isolation and contribute to the process of speciation itself. Therefore, the genetics of speciation can profit from studies of changes in gene expression as well as changes in gene frequency and genetic isolation ... A very large body of evidence shows that phenotypic novelty is largely reorganizational rather than a product of innovative genes. Even if reorganization was initiated by a mutation, a gene of major effect on regulation, selection would lead to genetic accommodation, that is, genetic change that follows, and is directed by, the reorganized condition of the phenotype.
>
> West-Eberhard, 2005, pp. 6543, 6547

The message here is that the very idea of what constitutes an 'environment' and of the relations between internal and external factors in evolution needs to be drastically revised. The constellation of phenomena we are going to see next confirms that.

Epigenetics and imprinting

Various processes of strictly local chemical alteration (technically called methylation, phosphorylation, acetylation, etc.) normally target the proteins that compose the histones, that is, the multiply repeated reels around which DNA is coiled in the chromosomes, forming what is called 'chromatin'. The similarity of wording with 'chromosome' is totally non-accidental, since chromosomes are formed by long uninterrupted strings of DNA annealed around the chromatin reels. The net effect of these ubiquitous and vital chemical modifications of the proteins forming the chromatin is one of physically exposing genes to the processing machinery of the cell, activating them, or burying them into the grooves of the chromatin, blocking their expression. Specific locations along the very long helix of the DNA itself can be the target of such micro-chemical modifications,[10] especially methylation, resulting in an additional and all-important mechanism of genetic regulation.[11]

Genes and possibly entire genomes are susceptible to many such minute, pervasive and crucial chemical modifications during development and then throughout adult life. The effect of such modifications of genes and genomes that are transmitted by the mother or by the father (especially in mammals) constitutes what is technically called parental 'imprinting' of the genes.

About 80 imprinted genes are presently known in the mouse and most tend to occur in clusters. Their activity is suppressed or silenced in a parent-specific pattern. Both maternal and paternal imprinting have been reported. Short sequences of RNA that do not code for proteins, in particular the gene regulatory factors we mentioned earlier, called micro-RNAs, are especially affected (Peters and Robson, 2008). Recent studies in humans of identical (monozygotic) and fraternal (dizygotic) twins have shown that the effect of sharing the same placental milieu adds to that of sharing the same genes (Kaminsky *et al.*, 2009). The effects of these chemical modifications of genes, giving different phenotypes in spite of an identical underlying genetic DNA, are actually intensely studied and have created a whole field of inquiry called 'epigenetics' (This already constitutes a vast scientific literature.

For a comprehensive treatment see Allis *et al.* [2006], for accessible early summaries see Pray [2004], and for more specific treatments see Rutherford and Henicoff [2003] and Vercelli [2004].)

The heritability of some such epigenetic modifications has been suggested in rodents, certainly down to the second generation, possibly as far as the fourth (Anway and Skinner, 2006; Jirtle and Skinner, 2007). In humans, probable cases of epigenetic inheritance of male infertility and susceptibility to other diseases have been reported. The clearest and most studied case of the heritability of epigenetic modifications in humans is offered by the tragic, Nazi-inflicted, Dutch famine in the winter of 1944–1945. The women who managed to give birth in those months had babies of height and weight considerably less than the norm. The surprising datum is that now, their granddaughters, in spite of their history of perfectly normal nutrition, still give birth to babies measuring below the norm. Another case showing that imprinted genes are good candidates to mediate nutrition-linked transgenerational effects on growth comes from Sweden. In this case there is also a paternal, probably sperm-mediated, imprinting (Pembrey, 2002). Analysing a long record of early nutritional influences on cardiovascular and diabetes-related mortality, Kaati, Bigren and Edvinsson of the University of Umeå exploited records of annual harvests from an isolated community in northern Sweden that go back as far as 1799 to explore the effects of food availability across three generations (Kaati *et al.*, 2002). In essence, scarcity of food in grandfather's periods of slow growth was associated with a significantly extended survival of the grandchildren for many years, while food abundance was associated with a greatly shortened lifespan of the grandchildren. Opposite effects have been attributed to the epigenetic role of nutrition for the grandmothers. The crucial steps in the process seem to have been the nutritional situation during the formation of sperm in the grandfathers, and the formation of the ovules in the grandmothers (Kaati *et al.*, 2002).[12]

A metaphor adopted by one of the pioneers in the field of epigenetics (Randy Jirtle of Duke University) suggests that the genome is the 'hardware', while epigenetic modifications are the 'software'.

Indeed, these modifications of the phenotype happen without any change in the corresponding DNA, much as the same hardware can be programmed to run different software.

Very recent data, from January 2009 (Tariq *et al.*, 2009) explain in detail the effects originally discovered by Waddington and confirm the earlier data and insights of John M. Rendel (see above). Chaperones such as HSP90 connect an epigenetic network, controlling major developmental and cellular pathways, with a system sensing external cues (heat shocks, for instance). Damages to this connecting process explain the manifestation of abnormal traits and the relatively rapid fixation of these under shock followed by selective breeding. Some 50 years later, the perplexity (not to say scepticism) with which Waddington's original experiments were received has no reason to persist. What he had aptly called 'genetic assimilation' and 'canalization' is now explained through complex, but perfectly mechanistic, molecular mechanisms.

The jury is still out on the heritability of epigenetic modifications, and we do not know yet how generalized and frequent epigenetic inheritance will turn out to be, and whether it will be ascertained to have hereditary effects even after the second generation down the line. As a result, especially in some popularization reports, this domain has raised perplexities (excessive, in our opinion), caused by the fear that Lamarckism may be making a comeback.[13]

Jumping genes and 'horizontal' gene transfer

We wish to insert into this summary sketch of non-Darwinian and non-selectionist interactions between organisms and different kinds of environments (including internal environments), a pervasive, evolutionarily crucial, phenomenon that has been unravelled in recent years: 'horizontal' gene transfer. This is an exchange of genes that takes place not in the canonical way, not by descent from one generation to the next (the transmission is then called 'vertical'), but rather between organisms that exist, so to speak, side by side (that's why it's called 'horizontal' transmission). Horizontal transmission takes place through a process that is not too dissimilar from the way viruses infect living cells. However, the case that is relevant to evolution is the frequent one in which these genetic 'outsiders' end up inserting themselves permanently into the genome of a species and then being transmitted 'vertically' in the canonical way through successive generations. First discovered in maize by Barbara McClintock (1902–1992) in the 1940s and 1950s (her work was much belatedly rewarded with a Nobel Prize in 1983), this phenomenon is ubiquitous. It has been estimated that at least 45 per cent of our genes derive from such horizontal gene transfers. These genetic elements are called in technical terms 'transposons' or transposable elements (TEs). TEs are present in virtually all species and often contribute a substantial fraction of their genome size, as they do in our own species. They can also 'transpose' or move around to different positions within the genome. A type of transposon called a retrotransposon is transcribed into RNA and then reintegrated into the genomic DNA, after its RNA is 'transcribed back' into DNA (a process that is the inverse of that we saw earlier). The most common form of retrotransposons in the human genome are the so-called Alu elements, which occur in more than one million copies in each of our genomes and occupy approximately 10 per cent of the whole human genetic sequence. They appear to be characteristic of primates and their possible role in hominization has been explored.[14]

A signal case is the full-blown immune system of mammals, called 'adaptive' because it is added on top of the so-called 'innate' immunity system already present in insects and because, as we all know, it 'adapts' to infections throughout our life. This extremely complex apparatus, comprising specific antigen receptors, immunoglobulins, B lymphocytes and T lymphocytes, seems to owe its evolutionary origins to the chance insertion of two transposons, called *RAG1* and *RAG2* (the latter possibly an ancient duplication of *RAG1*), into an ancestor of modern sharks, followed by a sort of multiplication explosion of those inserted genes, which were originally connected to the specific replication and multiplication of segments of DNA (Agrawal *et al.*, 1998; Hiom *et al.*, 1998; Malecek *et al.*, 2008). There is agreement that understanding the functional roles, evolution and population dynamics of TEs is essential to understanding genome evolution and function. These chance insertions represent, at the level of the genomic apparatus, rather massive injections of new genetic material, variably followed by further transpositions internal to the genome, internal stabilization, multiplication, partial loss and further mutations. These transposable elements represent a kind of genomic and evolutionary wild card, presently under intense scrutiny. How the different TEs in different species are subsequently subjected to internal and external selection remains to be defined.[15]

In the present context, we wish to point out that horizontal genetic transfer is the rule, rather than the exception, in microorganisms, to the point that the very notion of a 'tree of descent' is being questioned. The preferred metaphor today in microbiology is that of a bush or a network (Doolittle, 1999). The very idea of the tree has been questioned, perhaps too radically, also for the evolution and descent of higher animals. Be it as it may, the relatively massive, sudden and repeated 'horizontal' introduction of genetic material is a new factor in the present picture of evolution. It was worth being reported here, because we are ascertaining one more evolutionary process that was not contemplated in the standard neo-Darwinian model.

Interchangeability between reactions to external and internal perturbations

As we have learned from Mary-Jane West-Eberhard (see above), change in trait frequency and genetic accommodation of the threshold of expression of a novel trait frequently occur as consequences, rather than causes, of phenotypic change. The plasticity of adaptations to new external inputs may well be deployed also when heritable internal changes occur. In particular, developmental plasticity may be what allows new adaptive evolutionary changes to be stabilized, once they have occurred by mutation. Interestingly, it has been suggested that this effect may have been particularly strong in the evolution of the brain. For instance, Kirschner and Gerhart (2005) explain how shaving off whiskers in the mouse predictably affects 'backwards' the developmental pattern of organization of the nerve fibres and the brain centres to which they are connected.[16] This is the consequence of an external perturbation (the shaving of the whiskers). But the blind mole presents a perfectly analogous, although in this case congenital (i.e. internal) reorganization of its nervous system: the sensory tactile appendages of the nose 'project' to the barrels of the brain cortex in the same geometric pattern as do the mouse whiskers in an ordinary non-blind mouse whose whiskers have been shaved.

Another signal case of interchangeability of the effects of exogenous and endogenous causes is sex determination in reptiles. There are, generally speaking, two avenues to sex determination: genotypic sex determination (GSD), which is always the case in mammals, and temperature-dependent sex determination (TSD), frequent in other vertebrates (such as reptiles). Population density, food scarcity and differential sex-related mortality may also be factors (see Mary-Jane West-Eberhard's 2003 book for cases of alternating GSD and TSD in very close species of reptiles). The distribution of TSD and GSD across reptiles suggests several independent evolutionary transitions in sex-determining mechanisms, but transitional forms had yet to be demonstrated until one such case was recently observed in Central Australia.

The bearded dragon lizard (*Pogona vitticeps*) offers an instance of smooth transition between temperature-induced and gene-induced sex determination (Quinn *et al.*, 2007). Unlike mammals, in which the chromosomal structure of females is XX and that of males is XY, in this species ZZ individuals are males, while ZW individuals are females. There is a temperature-sensitive gene product present in two copies in males but in only one copy in females. The gene is fully active at intermediate temperatures but becomes progressively inactivated at higher temperatures. Reversal of the ZZ genotype to the female phenotype at extreme temperatures will bias the phenotypic sex ratio towards females and drive down the frequency of the W chromosome under frequency-dependent selection. This could account for the pattern observed in many TSD reptiles, in which both low and high temperatures produce 100% females, yet intermediate temperatures produce predominantly (occasionally 100%) males. (For still another clear case of interchangeability, see note 1 and figure 2).

We wish to stress, as Kirschner and Gerhart do, that the interchangeability of external factors and genetic fixation can well redraw the whole picture of evolution. The basic mechanisms are the same, and biologists are only beginning to understand the evolutionary impact of this fact.

Conclusions for this chapter

The internal constraints we reviewed in Chapter 2 were limitations to what *can*, in the first place, become the target of natural selection. The multiple levels of regulation and the genetic, developmental and evolutionary compartmentalization of organisms we reviewed in Chapter 3 constituted conditions on *how* natural selection can operate, if and when it does. Neo-Darwinists have tried to adjust to most of these processes by expanding its scope and invoking other kinds of natural selection, essentially multilevel adaptations and internal selections to internal milieus. Possibly, they may now want to try even harder and incorporate into their adaptive explanations processes such as those reviewed in this chapter: selection without adaptation, genetic assimilation, genotypic and phenotypic plasticity, contingency, sudden explosions of new forms, transposable elements, epigenetic regulations and the interchangeability of reactions to internal and external factors. These ought to increase, we think, the discomfort of classical neo-Darwinians. If they really want to go this far, then their theory will have to be subverted, not just reformed or expanded – a consummation devoutly to be wished. These are some of the tiles of a completely new and rapidly expanding mosaic, not additions to the old one. The following chapter will, we hope, accentuate the discomfort.

5

THE RETURN OF THE LAWS OF FORM

In the previous chapters we saw several constraints 'from below', from molecular interactions all the way up to phenotypes, and canalized interactions 'across' different kinds of internal organization and the milieus with which they interact. What we are going to see now is an entire spectrum of other factors that must have played a major role in evolution and that are equally, if not even more, alien to adaptation and natural selection. For historical reasons, and for want of a better term, we will call these 'the laws of form'. These are, in a sense, constraints 'from above',[1] because the mathematical and physico-chemical laws that explain spontaneous self-organization and the 'discovery' of optimal solutions exceed the boundaries of biology and are necessarily quite abstract. Lest we might be accused of wanting to 'reduce' biology to physics and chemistry, we wish to make it clear that these factors are intimately enmeshed with hosts of contingent evolutionary happenings and mis-happenings (from meteorites to glaciations, from volcanic eruptions to floods and much more) and with the manifold genetic and epigenetic contingent mechanisms we have just reviewed. The important emerging consideration is that these physico-chemical invariants also play a role in evolution, not that they are all there is to evolution. Of course not.

When very similar specific morphologies (Fibonacci series and

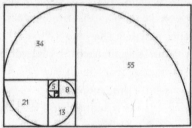

In the Fibonacci series, each term is equal to the sum of the two preceding ones (1, 1, 2, 3, 5, 8, 13, 21, 34, 55, and so on). By connecting, with a continuous curve, the outer vertices of an ordered pattern of juxtaposed squares that have areas given by the Fibonacci series we obtain the Fibonacci spiral (© Ian Stewart). The presence of Fibonacci patterns is ubiquitous in nature, from galaxies to seashells, from magnetized droplets in a viscous medium to the organization of florets in plants (as shown in the figure for sunflowers). Moreover, the numbers of such spirals (clockwise versus counterclockwise) are typically Fibonacci numbers (21 and 34 respectively).

Fibonacci spirals[2]) are observed in spiral nebulae, in the geometrical arrangement of magnetically charged droplets in a liquid surface, in seashells,[3] in the alternation of leaves on the stalks of plant stems and in the disposition of seeds in a sunflower, it can hardly be that natural selection is responsible.[4,5] The relative rates of growth and the initial stages in a plant's production of lateral buds (florets) is controlled by many hormones and networks of metabolites, proteins and so on that are under the control of genes and their regulatory networks. Yet, there is seemingly a simple 'rule', which is to place the new bud as far away as possible from the inhibitory hormone effects of the apex and of the last placed buds. Every species does this according to the genes and their regulatory networks that control the types, magnitudes of effects and amounts of inhibitory and stimulating hormones. The simple 'rule' is not, as such, dictated by the genes – it is something to which biological growth has to submit. It is the result of the laws of physics and chemistry creating constraints on possible biological forms, more particularly on stable and reproducible biological forms. This is what, basically, the expression 'laws of form' tries to capture.

As we are going to see in some paradigmatic instances, the relevant search space would be so huge that the hypothesis of such 'rules' and such constraints on stability having been found by blind trial and error followed by natural selection becomes exceedingly improbable. It is even harder to suppose that some genetic machinery is specifically (one has to insist on this: specifically) responsible for coding these forms *as such*. It's vastly more plausible to suppose that the causes of these forms are to be found in the elaborate self-organizing interactions between several components that are, indeed, coded for by genes (protein complexes, morphogenetic gradients, hormones, cell–cell interactions, and so on) and the strictures dictated by chemical and physical forces. The latter are vastly more ubiquitous and vastly less modular than biological processes. They transcend the biological subdivisions into species, genera, families, orders, classes and phyla. The vagaries of genetic and developmental factors operating over hundreds of millions of years, together with various internal and external levels of selection, must have been exploring the narrow channels allowed by maximization principles that are applicable to biology but exceed its boundaries.

In the apt words of mathematician Peter Timothy Saunders, someone who has been criticizing standard neo-Darwinism for many years and has insisted on the importance of the laws of form (Saunders, 1980), biologists have to delimit the realm of possible creatures first, and only then ask about natural selection: 'The primary task of the biologist is to discover the set of forms that are likely to appear ... [for] only then is it worth asking which of them will be selected' (Saunders, 1992, p. xii).

Another main advocate of the importance of the laws of form in biology, Stuart Kauffman, rightly (and somewhat sadly) says, in the preface to his important book on the origins of biological order: 'No body of thought incorporates self-organization into the weave of evolutionary theory' (Kauffman, 1993).

As we are going to see, there are good reasons for this divorce, although a recent and still somewhat sporadic return of the laws of form into biology may be conducive to some integration in the decades to come.

A little bit of history

The monumental pioneering work of D'Arcy Wentworth Thompson launched the very expression 'laws of form' in the early twentieth century (Thompson, 1917; reprinted and edited by Tyler Bonner in 1992).[6] He made the prescient suggestion that biologists had over-emphasized the role of evolution and underemphasized the roles of physical and mathematical laws in shaping the form and structure of living organisms.[7] In many ways his vast and ambitious work was premature, because the discovery of the biochemical and genetic bases of growth and form were still in the future, and because the mathematics mobilized to explain the phenomena was inadequate.

A few years later, in 1924, the Italian mathematician Vito Volterra (later summarized in a monograph in French [Volterra, 1931] and the American mathematician Alfred J. Lotka [Lotka, 1925, 1956]) independently and convergently discovered differential equations regulating the oscillatory equilibria of predators and prey in ecosystems, and applicable also to sustainable rates of growth, birth and mortality rates, biochemical cycles and rates of energy transformations, and even the evolution of human means of transportation and fluctuations in financial markets. These equations soon became staple food for mathematical ecologists and theoretical chemists the world over, and still are. Only very recently have some timid links with genetic networks been tried out.

The importation of the laws of form into biology proper had to wait several decades. In 1952, Alan Turing tried to explain biological patterns on the sole bases of canonical equations of chemical diffusion (Turing, 1952; Saunders, 1992). This long-forgotten paper had, in hindsight, major flaws,[8] and the development of molecular genetics ever since the late 1950s paid no attention to it (with the exception of the British geneticist and embryologist Conrad Hal Waddington, cited by Turing in his paper, whose work we have briefly reviewed in the previous chapter [Waddington, 1957], and to which we will return shortly).

In the meantime, the Russian chemists Boris P. Belousov and Anatol M. Zhabotinsky discovered the spontaneous formation of complex shapes and permanently oscillating reactions (spontaneously arising from perfectly homogeneous solutions).[9] The Russian-Belgian physical chemist Ilya Prigogine (1917–2003) later developed this whole domain of inquiry (labelled 'dissipative structures') into a high art, writing down the complete physical and chemical theory of these phenomena, down to quantum physics (his 1977 Nobel lecture remains illuminating; see Prigogine, 1993). Interestingly for us, a debate between Prigogine and Jacques Monod flared up, in which the role of natural selection versus the role of spontaneous morphogenesis was the cornerstone of the disagreement. Neglected or marginalized by Monod (just as it was by all his colleagues in molecular genetics) but touted by Prigogine, the importance of complex spontaneous morphogenesis in evolution had still to emerge into full view.

A prestigious ally of Prigogine's was the French mathematician

René Thom (see Prigogine, 1993), who had been awarded the coveted (by mathematicians the world over) Fields Medal in 1958 for the theory of structural stability and morphogenesis. His universal classification of discontinuous morphogenetic forms into seven elementary 'catastrophes' under even slight critical variations of the control parameters[10] prompted Thom to venture into possible (and quite unfortunate) applications well beyond biology (sociology, psychoanalysis, semantics, etc.). The English translation of Thom's main synthetic treatise *Structural Stability and Morphogenesis* (French original 1972, English edition 1975; see Thom, 1975) has, significantly, a foreword by Conrad Hal Waddington, who was in those years looked upon with some suspicion – or simply ignored – by mainstream geneticists and embryologists, although in recent years his early discoveries of the role of epigenetics have vindicated the importance of a lot of his data (see the previous chapter). Waddington coined terms such as 'canalization', 'canalized selection', 'chreod' and 'homeorhesis' to capture the subtle interaction of morphogenetic processes under the influence of genes in getting around or exploiting the constraints imposed by physical and geometric factors acting on embryology and evolution. Waddington was held in great suspicion by mainstream molecular geneticists in those years (approximately the 1960s and 1970s).[11]

This brief history can be wrapped up by mentioning later contributions[12] by Stuart Kauffman, Brian Goodwin, Lewis Wolpert, Antonio Lima-de-Faria, Antonio Garcia-Bellido, Stuart Newman and Gerd Mueller, among others.[13] Special mention needs to be made of the recurrent insistence on the significance of laws of form in biological evolution by the late Stephen Jay Gould and his colleague and co-author Richard C. Lewontin.[14] This whole field has been widely ignored by entire generations of militant geneticists, 'wet' molecular biologists and molecular embryologists.[15] The age of specificity, starting with the discovery of the structure of DNA by Crick and Watson in 1953, steered molecular biology away from these relatively generic approaches (Watson and Crick, 1953). Perhaps for that reason, no concrete problem in molecular genetics or microbiology has yet been solved by appeal to laws of form, although connections between these fields are proliferating. As we are going to see, in diverse quarters, somewhat episodically, there is a return of the laws of form into biology. It's reasonable to expect more and more in the years to come. The phenomena that have been uncovered represent serious and diversified challenges to gradualistic adaptationism and neo-Darwinism.

The 'fourth dimension' of living systems

The body masses of living organisms vary between 10^{-13} grams (bacteria) to 10^8 grams (whales), that is, by 21 orders of magnitude. It's interesting to see how other physico-chemical and biological properties and processes, and their ratios, scale with mass. How, for instance, surfaces and internal rates of transport, rates of cellular metabolism, whole organism metabolic rate, heartbeat, blood circulation time and overall lifespan scale with mass. These are, of course, all three-dimensional systems, so it seems astounding that all the scaling factors, encompassing microorganisms, plants and animals, are multiples of a quarter, not of a third.[16]

The puzzle has been solved in collaborative work by physicists and biologists at Los Alamos, Santa Fe and Albuquerque. In essence, they have discovered a 'fourth dimension' of biological systems. The explanation of the one-quarter scaling laws was found 'in the fractal-like architecture of the hierarchical branching vascular networks that distribute resources within organisms' (West et al., 1999, p. 1677). Their papers reveal a remarkable convergence between the experimental values and the predicted ones (sometimes down to the third decimal), under this hypothesis of fractal-like architecture, for properties such as radius, pressure and blood velocity in the aorta; cardiac frequency; number and density of capillaries; overall metabolic rate; and many more. Their mathematically detailed model (refined over the years) (West et al., 2002) takes into account biological data such as the 60,000 miles of the entire circulatory system of a human body (capillaries notably included) and the fact that the diameter of capillaries is an invariant in the realm of vertebrates.

Guiding criteria have been the maximization of the inner and outer exchange surfaces, while minimizing distances of internal transport (thus maximizing the rates of transport). A passage in the 1999 paper deserves to be quoted in full:

> Unlike the genetic code, which has evolved only once in the history of life, fractal-like distribution networks that confer an additional effective fourth dimension have originated many times. Examples include extensive surface areas of leaves, gills, lungs, guts, kidneys, chloroplasts, and mitochondria, the whole-organism branching architecture of trees, sponges, hydrozoans, and crinoids, and the treelike networks of diverse respiratory and circulatory systems … Although living things occupy a three-dimensional space, their internal physiology and anatomy operate as if they were four-dimensional. Quarter-power scaling laws are perhaps as universal and as uniquely biological as the biochemical pathways of metabolism, the structure and function of the genetic code and the process of natural selection.
>
> West *et al.*, 1999, p. 1679

In the words of these authors, natural selection has 'exploited variations on this fractal theme to produce the incredible variety of biological form and function', but there were 'severe geometric and physical constraints on metabolic processes'.

The conclusion here is inescapable, that the driving force for these invariant scaling laws cannot have been natural selection. It's inconceivable that so many different organisms, spanning different kingdoms and phyla, may have blindly 'tried' all sorts of power laws and that only those that have by chance 'discovered' the one-quarter power law reproduced and thrived. The maximization principles that have constrained such a bewildering variety of biological forms are of a physico-chemical and topological nature. Biochemical pathways, the genetic code, developmental pathways and (yes) natural selection cannot possibly have shaped these geometries. They had no 'choice' (so to speak) but to exploit these constraints and be channelled by them.

The same kind of lesson comes from calculations, and data, in the domain of brain connectivity.

Non-genomic nativism

The expression 'non-genomic nativism' was coined by Christopher Cherniak and collaborators at the University of Maryland in 1999 and has been used by them ever since (Cherniak *et al.*, 1999; Cherniak *et al.*, 2004; Cherniak, 2009). Combining a detailed anatomo-physiological analysis of the nervous system of the nematode, all the way up to the cortex of cats and monkeys, with a long series of computational simulations, it emerged that the minimization of connection costs among interconnected components appears either perfect, or as good as can be detected with current methods. Such wiring minimization can be observed at various levels of nervous systems – invertebrate and vertebrate – from placement of the entire brain in the body down to the subcellular level of neuron arbor geometry. These instances of optimized neuroanatomy include candidates for some of the most complex biological structures known to be derivable 'for free, directly from physics' [*sic*]. Such a 'physics suffices' picture for some biological self-organization directs attention to innate structure via non-genomic mechanisms.

Since general network optimization problems are easy to state, but enormously computationally costly to solve exactly (they are in general what computer scientists call 'NP-hard': that is, exponentially exploding in complexity), some simplifications had to be introduced. A 'formalism of scarcity' of interconnections (the so-called 'Steiner trees') was borrowed from engineering and used as the computational engine of network optimization theory, which characterizes efficient use of limited connection resources.[17] Cherniak *et al.* conclude that the cortex is better designed than the best industrial microchip. For the macaque, fewer than one in a million of all alternative layouts conform to the adjacency rule better than the actual layout of the complete macaque set. In the relatively simpler case of the nematode *Caenorhabditis elegans*, its nervous system having been the first ever to be fully mapped, the actual layout of eleven ganglia is the wire-length-minimizing one, out of 40 million alternative possibilities.

In a 2009 paper, Cherniak specifies that:

> The neural optimization paradigm is a structuralist position, postulating innate abstract internal structure – as opposed to an empty-organism blank-slate account, without structure built into the hardware (structure is instead vacuumed up from input). The optimization account is thereby related to Continental rationalism; but for brain structure, rather than the more familiar mental structure.
>
> Cherniak, 2009

His message is that there is a 'pre-formatting' issue for evolutionary theory. Seeing neuroanatomy so intimately meshed with the computational order of the Universe brings one back, as he suggests, to the explanatory project of D'Arcy Wentworth Thompson and Turing (Cherniak, 2009). There is, indeed, in our terminology, a return of the laws of form.[18]

Further examples

These further examples all share the property that we have emphasized: evolution seems to have achieved near optimal answers to questions which, if pursued by the application of exogenous filters to solutions generated at random, as the neo-Darwinist model requires, would have imposed searching implausibly large of spaces of candidate solutions.[19] This seems an intractable enigma, unless prior filtering by endogenous constraints is assumed.[20]

The brain's grey and white matter

The segregation of the brain into grey and white matter has been shown by biophysicists to be a natural consequence of minimizing conduction delay in a highly interconnected neuronal network. A model relating the optimal brain design to the basic parameters of the network, such as the numbers of neurons and connections between them, as well as wire diameters, makes testable predictions all of which are confirmed by anatomical data on the mammalian neocortex and neostriatum, the avian telencephalon and the spinal cord in a variety of species (of mammals and birds) (Wen and Chklovskii, 2005).

Invariants of animal locomotion

Scaling laws and invariants in animal locomotion have been uncovered by engineer Adrian Bejan (Duke University) and biologist James H. Marden (University of Pennsylvania) by considering that 'animal locomotion is no different than other flows, animate and inanimate: they all develop (morph, evolve) architecture in space and time (self-organization, self-optimization), so that they optimize the flow of material' (Bejan and Marden, 2006, p. 246).[21]

Pulling together, in their model, 'constructal' (*sic*) principles, equally applicable to the morphing of river basins, atmospheric circulation, the design of ships and submarines, and animal locomotion, regardless of whether it consists in crawling, running, swimming or flying, they can explain the nature of the constraints and derive principles for optimized locomotion. The parameters that characterize, for each species, the locomotion that accomplishes the most for unit of energy consumed, that is, the points at the bottom of the U-shaped curve of cost versus speed, align neatly along a straight line in a logarithmic scale. Plotting optimal force against body mass, from the smallest marine creature to elephants, this straight line scales the very narrow range of speeds that maximize, for each species, the ratio of distance travelled to energy expended.

Simple equations that correlate body mass, body density, body length, the gravitational acceleration and the coefficient of friction reveal that even the distinction between flying, swimming and walking (crawling, running) is immaterial. Physical principles of optimization and simple scaling laws govern the phenomena of animal locomotion.

The physics of birdsong

Two physicists and a biologist, publishing in a physics journal, show that

> The respiratory patterns of the highly complex and variable temporal organization of song in the canary (*Serinus canaria*) can be generated as solutions of a simple model describing the integration between song control and respiratory centres. This example suggests that sub-harmonic behaviour can play an important role in providing a complex variety of responses with minimal neural substrate.
>
> Trevisan et al., 2006

A straightforward generalization to other kinds of birdsong in other species of singing birds is plausibly anticipated.

We want to raise the issue: have all sorts of suboptimal neuronal setups and of the ensuing suboptimal singing patterns been tried out at random over the aeons and natural selection made it so that only the optimal singers left descendants? Did the sub-harmonic equations become slowly encoded in the canary genes by chance trials and selection? Or are we witnessing an instance of physical optimization constraints channelling genetic, developmental and behavioural traits? Nobody at present has an idea of how these physical optimization factors interact, over evolutionary times and over ontogenetic times, with the genetic and epigenetic machinery of organisms. This is why the younger generation of biologists and biophysicists will enjoy unexpected discoveries in the decades to come. Lucky them!

The perfect leaves

Regarding the plant kingdom, a team of American and French biologists and physicists has recently determined by means of mathematical equations and artificial simulations that the spontaneous biological 'design' of leaves is perfect. They generated parallel networks of channels in layers of artificial polymeric material (large molecules formed by multiply repeated smaller molecules chained together), and showed through such simple networks that the scaling relations for the liquid flow driven by evaporation reveal basic design principles that satisfy the most stringent engineering requirements for devices that are driven by evaporation–permeation factors. These authors highlight the role of physical constraints on the biological design of leaves (Noblin *et al.*, 2008).[22] They show that the flow rate through their biomimetic and real leaves increases linearly with channel density until the distance between channels is comparable with the thickness of the polymer layer, above which the flow rate saturates. A comparison with plant vascular networks shows that the same optimization criterion can be used to describe the placement of veins in leaves.

Optimal foraging strategies:[23] the honeybees

As von Frisch had taught us, at the start of a foraging period some individual honeybees go out foraging on their own ('proactive' searchers) and some ('reactive' searchers) stay in the hive awaiting information from returning foragers that is conveyed by the famous 'bee dance' (von Frisch, 1967). The issue to be solved was: which optimal percentage of individuals should go out and forage on their own and which correspondingly optimal percentage should wait for information? Clearly, it can't be the case that all searchers are reactive; so the question arises whether there is an optimal percentage of proactive to reactive searchers (as a function of colony size and the availability of perishable food). Researchers (Dechaume-Moncharmont *et al.*, 2005) combined measurements of actual foraging behaviours with a mathematical model of the energy gain by a colony as a function both of the probability of finding food sources and of the duration of their availability. The key factor is the ratio of proactive foragers to reactive foragers. Under specifiable conditions, the optimum strategy is a totally independent (proactive) foraging for all the bees, because potentially valuable information that reactive foragers may gain from successful foragers is not worth waiting for. This counter-intuitive outcome is remarkably robust over a wide range of parameters. It occurs because food sources are only available for a limited period. But their study emphasizes the importance of time constraints and the analysis of dynamics, not just steady states, to understand social insect foraging. The predictions of their model for optimal foraging, often quite counter-intuitive, have been confirmed both in the wild and in laboratory conditions (Dechaume-Moncharmont *et al.*, 2005). The bees appear to be 'sitting' (so to speak) at the optimum of the curve of the possible ratios of proactive versus reactive foragers in a variety of situations.

Once again, we want to raise some key issues for the theory of evolution: it's not possible that all sorts of foraging strategies have been tried out at random over the aeons, and that natural selection determined that only the optimal foraging bees left descendants. Maybe no neo-Darwinian wishes today to suggest this kind of crude hypothesis. A somewhat more plausible picture is that, once some change in foraging strategies has occurred, the range of further changes beyond that will have changed completely. The subset of possible further small changes around the present behavioural phenotype is constrained, and so is the subset around the underlying genotype and the developmental pathway. As suggested by Richard Lewontin, the metaphor is rather one of finding one's way through a maze, with no possibility of wandering back to the starting point. The population, or the species as a whole, is committed to certain downstream passages. Every evolutionary change constrains to some subset, and new sub-subsets of possible further mutational effects at the next step. It's hard, at present, to go beyond such metaphors.[24] However, the picture of a blind search winnowed by selection is utterly implausible. Multiple stepwise canalization of variants, under the kinds of physical-computational constraints suggested by Cherniak *et al.* must have eventually led to an inbuilt computation of the optimal ratio of proactive and reactive foragers, somehow encoded in the interaction between genes, development and the action of some laws of form. The question here involves multiple individuals and their behaviour, and the solution will in due time turn out to be more complex than that of the individual canaries. Once again, nobody today really has a clue to a solution of these problems. These issues need to be raised nonetheless.

We have seen examples where it seems that only physico-chemical and geometric constraints can explain the narrow canalizations that natural selection must have explored. The case of the bees, and two more that we are going to see (just a sample among many more in the recent literature) are such that, once more, the space of possible solutions to be explored seems too gigantic to have been explored by blind trial and error. The inference appears to be that a highly constrained search must have taken place. Accordingly, the role of natural selection may have been mostly just fine-tuning. Or less.

The perfect wing stroke

The utility of a sixth or a fifth of a wing has been questioned for quite some time (including by one of us in past writing) as a challenge for gradualist adaptationism. With a sixth of a wing an animal does not fly a sixth of the time or a sixth of the distance. It does not fly at all. Therefore, the challenge for gradualist adaptationism is to explain how mutations capable of producing full wings can have accumulated silently over a long evolutionary time in the absence of any adaptive advantage. This issue was raised for the evolution of insect wings by Kingsolver and Koehl (1985). Insect wings are an evolutionarily significant novelty whose origin is not recorded in the fossil record. Insects with fully developed wings capable of flight appear in the fossil record in the upper Carboniferous (ca. 320 million years ago), by which time they had already diversified into more than ten orders, at least three of which are still extant. On the contrary, wingless insects are observed in the fossil record as early as the Silurian (ca. 400 million years ago) (Engel and Grimaldi, 2004). The intervening fossil record is poor, and no fossils showing intermediate stages in the evolution of wings have been identified (Yang, 2001). Several evolutionary scenarios have been proposed (Jockush and Ober, 2004). The interesting suggestion by Kingsolver and Koehl was that the initial selective factor had been thermal exchange, not mobility. A sixth of a wing in an insect cannot generate any lift, so no airborne mobility. That comes only after the wing reaches about 85 per cent of its full final size. But it does generate, by means of the tiny wing veins, a non-negligible capacity for raising the insect's temperature when exposed to sunlight. Their study was based on the comparative anatomy and physiology of several species of insects very similar to one another, except for having or not having wings, on calculations and on experiments with scale models in wind tunnels. Kingsolver and Koehl concluded that wings must have initially developed in insects as thermal organs, and later on, after gradual increase driven by this function, once a size capable of providing lift had been developed, further selection of a quite different kind ensued: selection for flight. If true, this would be another signal case of the role of free-riding traits (spandrels) in evolution, a topic to which we will return in a later chapter. In the meantime, however, the picture has somewhat changed. A complex interplay between master genes (homeotic genes) and tandem duplications of dorsal thoracic genes seems to be involved in the appearance of wings in insects (Carroll, 1995). It is still being debated whether insect wings are an evolutionarily sudden novelty, or have evolved by modification of limb branches that were present in ancestral arthropods (Jockush and Ober, 2004). Possibly, the 'exaptationist' process suggested by Kingsolver and Koehl and internal genetic changes may have acted in concert. We saw in the previous chapter that such interchanges between internal genetic reorganizations and external factors do actually take place.

A different tack, this time for the evolution of wings in birds, was taken in a paper published in *Nature* in January 2008 by Kenneth P. Dial, Brandon E. Jackson and Paolo Segre (Dial *et al.*, 2008). They present the first comparison of wing-stroke kinematics of the primary locomotor modes (descending flight and incline flap-running) that lead to level-flapping flight in juvenile ground birds throughout development. They offer results 'that are contrary both to popular perception and inferences from other studies'. Prior to this study, no empirical data existed on wing-stroke dynamics in an experimental evolutionary context. In a nutshell, starting shortly after hatching and continuing through adulthood, ground birds use a wing stroke confined to a narrow range of less than 20 degrees, when referenced to gravity, that directs aerodynamic forces about 40 degrees above horizontal, permitting a 180-degree range in the direction of travel. Estimated force orientations from the birds' conserved wing stroke are limited to a narrow wedge. A main result of their extremely detailed comparative analysis of the wing-stroke plane angle, estimated force orientation and angle of attack among locomotor styles is that, when wing-stroke plane angles are viewed side by side in both the vertebral and gravitational frames of reference, the wing stroke is nearly invariant relative to gravity, whereas the body axis reorients among different modes of locomotion.

Their experimental observations reveal that birds move their 'proto-wings', and their fully developed wings, through a stereotypic or fundamental kinematic pathway so that they may flap-run over obstacles, control descending flight and ultimately perform level flapping flight. Interestingly, these authors offer the hypothesis

> that the transitional stages leading to the evolution of avian flight
> correspond both behaviourally and morphologically to the tran-
> sitional stages observed in ontogenetic forms. Specifically, from
> flightless hatchlings to flight-capable juveniles, many ground birds
> express a 'transitional wing' during development that is represent-
> ative of evolutionary transitional forms.
>
> Dial et al., 2008

They say that locomotor abilities of extinct taxa, such as the recently discovered fossil forms possessing what is assumed to be 'half a wing', and long cursorial legs, might be better understood if we evaluate how proto-wings and hindlimbs function during ontogeny in extant taxa. Their experimental observations show that proto-wings moving through a stereotypic and conserved wing stroke have immediate aerodynamic function, and that transitioning to powered flapping flight is limited by the relative size of the wing and muscle power, rather than development of a complex repertoire of wingbeat kinematics.

Fine, but then, in our view, another problem arises for gradualistic adaptationism, because another kind of discontinuity is appealed to. In their own words: '... the gravity based wing-stroke did not come about through a long series of migrational stages of the forelimb (from ventro-lateral to lateral to dorso-lateral): rather, the primitive wing-stroke started in a similar orientation as we see it today in hatchlings using their proto-wings' (Dial *et al.*, 2008).

The angles of effective wing stroke are extremely narrow, as these authors have determined, and one wants to question the process through which this narrow wedge of angles became fixated even before there was any real flight. The amplitude of the search space for the optimal angle seems to be even more daunting than that of the search space for the series of migrational stages (ventro-lateral to lateral to dorso-ventral). One cannot help wondering, in this case too, whether physical (gravitational, aerodynamic) constraints have not narrowed down the morphological search space drastically. Evo-devo mechanisms seem once more to have been severely constrained by non-biological, and surely non-selectional, factors.

The zombifying wasp

Finally, a case (again, among many) in which the genetic programming of a complex behaviour leaves no doubt. Such behaviours can be shown to be completely automatic through the whole sequence, and unlearned. To cut a long story short, a particular species of wasp (*Ampulex compressa*) uses a venom cocktail to manipulate the behaviour of its cockroach prey. As in some other species of solitary wasp, the female wasp paralyzes the cockroach without killing it, and then transports it into her nest and deposits her eggs into the belly of the cockroach, so that the hatchlings can feed on the cockroach's live body. What is peculiar to this species of wasp is that, by means of two consecutive stings, separated by a rather precise time interval, in two different and precisely chosen parts of the cockroach nervous system, the wasp becomes capable of literally 'driving' the zombified cockroach into her prepared nest. The wasp does not have to physically drag the cockroach into the pit, because it can manipulate the cockroach's antennae, or literally ride on top of it, steering it as if it were a dog by a leash, or a horse by a bridle (Libersat, 2003). The first sting in the thorax causes a transient front leg paralysis lasting a few minutes. Some behaviours are blocked but not others. The second sting, several minutes after, is directly in the head.

As a result the wasp can grab one of the cockroach's antennae and walk to a suitable oviposition location. The cockroach follows the wasp in a docile manner like a dog on a leash (Williams, 1942; Fouad *et al.*, 1994). A few days later, the cockroach serves as an immobilized and fresh food source for the wasp's offspring.

Some evolutionary questions

This rather horrendous entomological saga suggests some key evolutionary questions. Such complex, sequential, rigidly pre-programmed behaviour could have gone wrong in many ways, at any one of its steps. The biochemical nature of the cocktail of venoms could have been different in many ways, being, as a result, either totally ineffective, or overdoing it, by killing the prey. The timing and location of the stings could have gone wrong in many ways, letting the cockroach recover, for instance, and kill the much smaller wasp. The wasp could have failed to 'understand' that the prey can be led by the leash, after these two master strokes, and could have painfully dragged the rather big body to the nest. And so on and so on. The ways in which this behavioural sequence could have gone awry are indeed innumerable. Not even the most committed adaptationist neo-Darwinians suppose that all kinds of alternatives have been blindly tried out by the ancestors of the wasp and that better and better solutions were progressively selected, and that this optimal solution was finally retained and encoded for in the genes. True: wasps have been around for a very long time (some 400 million years, maybe more) but even this is not a long enough time to try out innumerable alternative behavioural solutions, with alternative possibilities conceivable at each step of the behavioural sequence. What, then? No one knows at present. Such cases of elaborate innate behavioural programs (spider webs, bee foraging as we saw above, and many more) cannot be accounted for by means of optimizing physico-chemical or geometric factors. But they can hardly be accounted for by gradualistic adaptation either. It's fair to acknowledge that, although we bet that some naturalistic explanation will one day be found, we have no such explanation at present. And if we insist that natural selection is the only way to try, we will never have one.

A lot of tunnel at the end of the light

There is only so much that the 24,000 or so genes in the human genome can do to assemble a human being. To be sure, there are multiple gene regulations and networks of interactions, and morphogenetic attractors, and epigenetic modifications, and alternative gene splicing and complex interactions with many environments. The latter, in the case of humans, comprise 'culture'. The task of those relatively few genes looks awesome, even considering the remarkable 'multiplier' represented by alternative splicing (see Chapter 3). Among other complex structures, tens of millions of kinds of antibodies have to be produced, and 10^{11} neurons and 10^{13} situated synapses to be developed and fixated, and about 60,000 miles of veins, arteries and capillaries to be exactly placed in each of our bodies. Many processes of spontaneous self-organization surely take place, at many levels. That's where the physics of collective phenomena leaves a signature, in ways that still elude us. Both the genetic-epigenetic processes and these other processes are internally caused. Cherniak's notion of 'non-genomic nativism' (Cherniak *et al.*, 1999, 2004; Cherniak, 2009) is appropriate and, it appears, inescapable.

Wrap-up of this chapter

Neo-Darwinists are keen to say that natural selection never optimizes, it only finds locally satisfactory solutions.[25] From François Jacob's evolutionary 'tinkering' to Maynard Smith's and Dennett's 'satisficing',[26] emphasis is always put on this consideration. It is important to our critique of neo-Darwinism that the problem of finding optimal solutions to evolutionary problems by filtering candidates generated at random would often be intractable. But, as we have just seen, there are *some* instances of optimal (or near-optimal) solutions to problems in biology; so, if natural selection cannot optimize, then something else must be involved. Very plausibly, the 'something else' includes: physics, chemistry, autocatalytic processes, dissipative structures and principles of self-organization, and surely other factors that the progress of science will in due time reveal.

The moral here is a sort of dilemma for neo-Darwinists: even if we suppose, for the sake of argument, that natural selection does operate in the way that canonical neo-Darwinism claims it does, its degrees of freedom must be severely restricted. The minute proportion of the theoretical morpho-spaces of life that are actually occupied (McKinney and McGhee, 2003; McGhee, 2007; Raup, 1966) is something for which the theory of natural selection cannot account.[27] We are understandably awed by the variety and the diversity of the forms of life, but it is important to underline that, at an abstract level, when plotting the continuous possible variation of parameters of form, extant and extinct forms of life are a very tiny subset of what is possible in the abstract.[28]

As we just said, this dilemma would hold even if the theory of natural selection were otherwise basically correct. But the issue is arguably academic, as the following chapters in Part two will show that it isn't.

PART TWO

THE CONCEPTUAL
SITUATION

6

MANY ARE CALLED
BUT FEW ARE CHOSEN:
THE PROBLEM OF
'SELECTION-FOR'

'Population thinking' is currently the preferred way to construe the goal of evolutionary theories: they are meant to articulate principles according to which the distribution of phenotypic traits in a population changes over time (typically as the effect of causal interactions with ecological variables). This book is an extended argument that, if that's what a theory of evolution is supposed to do, then the theory that evolution proceeds by a process of natural selection cannot be true.

This chapter begins to spell out the argument. There are several sections. The first offers considerations showing that, at best, natural selection can't be the *whole* story about how phenotypes evolve. In fact, as we read the current literature and as we've seen in Part one, that isn't seriously in dispute these days. Perhaps it never was: certainly Darwin himself didn't think that natural selection could be the only mechanism of evolution. But he pretty clearly did think that natural selection is a very substantial part of the story; and that the slack would be taken up by adding details that elaborate, but don't conflict with, the basic selectionist model. Darwin thought that a refined theory of evolution would be a sophisticated adaptationism. As far as we can tell, that remains the consensus view.

The next several sections of the chapter introduce the problem of 'selection-for';[1] roughly, this is the question of which of an organism's phenotypic traits is causally responsible for some or other variation in its fitness. We'll review a number of what we take to be avatars of this problem, drawn not just from evolutionary theory but also from philosophy, psychology and semantics. Later sections are about what we take to be the similarity between these examples, and what they share with the objections we want to raise against adaptationism. This will mean having a sustained look at how selectionist explanations work: what a philosopher might call their 'logic'. We'll have a lot to say about how they are related to teleological and mechanistic explanations, and, above all, to intentional explanations. It's increasingly clear that these sorts of matters are central to any serious evaluation of adaptationism.

'Free-riders'

In 1979, Gould and Lewontin (Gould and Lewontin, 1979) wrote a paper that has since (rightly) become iconic. It's a good place for a discussion of selectionist accounts of evolution to start.

As we understand it, one of the main points of the Gould–Lewontin paper was to suggest a way in which there could be exceptions to the thesis that phenotypic traits are adaptations; exceptions, that is, to the thesis that (over the long run and unsystematic variables aside) phenotypic traits evolve in the direction of increased fitness. This issue is closely linked to the theory that natural selection is the main cause of phenotypic evolution, as selection for fitness is supposed to explain why phenotypes become increasingly well adapted over time.[2] Conversely, if there are phenotypic traits that don't augment fitness, they are prima facie counterexamples to the generality of natural selection. Gould and Lewontin examine a number of ways in which such cases might occur; 'free-riding' is one of these.

We are invited to consider the relation between arches and

spandrels in the design of cathedrals that have domes such as St Mark's in Venice, illustrated above. (Spandrels are the little triangles formed by the convergence of the arches that support a dome.) It's unproblematic why domed churches have arches: if they didn't, the domes would fall down. So, it is reasonable to think of arches using the analogy of adaptations; like adaptations, they're explained tele-ologically, by reference to the function they perform: the arches are there to hold the roof up.

Churches that have arches regularly have spandrels too; that being so, it's in the spirit of Darwinist theorizing to suppose that, just as there is something that the arches are there for, perhaps there is something that the spandrels are there for too. One is thus led to ask: What is it about churches with domes that spandrels are adaptations to? As with the giraffe's long neck and the polar bear's white coat and the elephant's large ears, one can imagine various hypotheses, of varying degrees of face plausibility: perhaps the spandrels are there to provide interesting spaces for muralists to decorate; or they're there to produce an appearance of perspective, or whatever. Gould and

Lewontin's point is that, as a matter of fact, all such adaptationist theories about spandrels are false; they are 'just so' stories, elaborated post hoc, to license teleological explanations of what are, in reality, functionless facts. Spandrels don't actually do anything, they are mere 'free-riders': geometry guarantees that if you choose arches, you get spandrels willy-nilly as a by-product. It's the arches that architects select for. The spandrels just come along for the ride.

The Gould and Lewontin (1979) paper (the full title of which is 'The spandrels of San Marco and the Panglossian paradigm: a critique of the adaptationist programme') raises many considerations that are germane to this book. For one thing, Gould and Lewontin fault the adaptationist programme:

> ... for its failure to distinguish current utility [of a phenotypic trait] from reasons of origin ...; for its unwillingness to consider alternatives to adaptive stories; for its reliance upon plausibility alone as a criterion for accepting speculative tales; and for its failure to consider adequately such competing themes as random fixation of alleles, production of non-adaptive structures by developmental correlation with selected features (allometry, pleiotropy, material compensation, mechanically forced correlation), the separability of adaptation and selection, multiple adaptive peaks, and current utility as an epiphenomenon of non-adaptive structures.
>
> Gould and Lewontin, 1979

They then proceed to dismantle, specifically and persuasively, some adaptationist 'just so' stories that were current at the time. They conclude with the warning that 'one must not confuse the fact that a structure is used in some way ... with the primary evolutionary reason for its existence and conformation.'

We greatly admire Gould and Lewontin's paper for its clarity and cogency, but we are nonetheless perplexed by some of the things the authors say. Thus, having put the reader on guard against simplistic and trivialized interpretations of Darwin, Gould and Lewontin declare their allegiance to selection as 'the most important of evolutionary mechanisms'. Not the *only* such mechanism, as Darwin himself had acknowledged, but still, according to Gould and Lewontin, the most important one. If that is the intended moral, then what Gould and Lewontin proposed, although it was widely viewed as extremely tendentious, is really a rather conservative amendment of the adaptationist tradition. They don't deny that natural selection is the central story about the evolution of many, possibly most, phenotypic traits. Moreover, despite their emphasis on 'laws of form', they don't deny that selection is the effect of exogenous variables. Churches have spandrels because their architects designed them to have arches, on which the spandrels free-rode. Giraffes have long necks because when relatively long-necked giraffes competed with relatively short-necked giraffes, ecological (hence exogenous) variables favoured the former. The salient difference is just that, whereas nobody intended or planned for long-necked giraffes, in the arch/spandrel case, the exogenous variables included the intentions of the architect. In short, both the presence of the spandrels and the length of the giraffe's neck are treated either as adaptations or as free-riders on adaptations. What Gould and Lewontin have on offer is a very sophisticated kind of adaptationism, but it's a kind of adaptationism all the same.

We think that Gould and Lewontin were right to postulate mechanisms of evolution whose operations can produce functionless phenotypic traits. But we also think that the deficiencies of adaptationism run a great deal deeper than Gould and Lewontin supposed. Actually, what is most problematic (so we're about to argue) is something that Gould and Lewontin cautiously endorse, and that Darwin announced frequently and explicitly in *The Origin of Species*: that *artificial* selection (in the Gould–Lewontin case, the selection of designs by architects; in *Origin*, the selection of phenotypes by breeders) is an appropriate model for *natural* selection. Adaptationists often say that this is just a harmless exegetical metaphor, but we're going to argue, to the contrary, that the putative analogy to artificial selection actually bears the whole weight of adaptationism. It's much like the arches and the domes; take the one away and the other collapses. This is intended to be a much more radical critique of the adaptationist programme than the one offered by Gould and Lewontin. If we're right, there is something wrong at the core of adaptationism that just fleshing it out with caveats won't fix.

However, it's best to approach the exposition indirectly. That's because the kinds of points that our argument turns on first emerged in discussions of topics that seem, at first glance, to be a considerable distance from the evaluation of adaptationism – and, indeed, from one another. We're interested in showing how they all connect. We ask the reader's indulgence while we set out the various pieces. We promise to put them together before this chapter ends.

Avatars of the 'selection-for' problem: teleology[3]

People have wondered, from time to time, what the heart is for. The informed answer is, of course, that it's there to pump the blood. Our present interest is: What, if anything, makes that the right answer?

It was, to our knowledge, C. G. Hempel (1965) who first pointed out the following worry: no doubt the heart pumps blood, but it also does a lot of other things. For example, the heart makes heart noises; and presumably, it couldn't pump blood if it didn't. So why is pumping blood, rather than making noise, the function of the heart?

The reader will have noticed that there is a certain symmetry between Hempel's puzzle about the heart's function and the Gould–Lewontin puzzle about why arches are adaptations but spandrels aren't. This is, of course, not a coincidence. One thing the two questions have in common is that it's plausible that both turn on the status of *counterfactuals* (that is, of hypothetical propositions with false antecedents: *what would have been the case if* …). For instance, if the architect *could have* arranged to have the arches without the spandrels, he might well have chosen to do so. If Mother Nature[4] could have arranged to get the blood pumped without making the heart noise, she might well have preferred that. But heart noises, like spandrels, are forced options (that is, they are both free-riders): it's only because of constraints that the plumbing imposes on blood pumps that we have hearts that make the noises that they do. It's only because of constraints that geometry imposes on the support of domes that churches have spandrels when they have arches. And so forth. This pattern will recur as the discussion proceeds: it is entirely character-istic of what we're calling 'selection-for problems' that relevant *coun-ter*factuals are what decide between two (or more) hypotheses that are equally compatible with the *actual* data. So, in the Gould–Lewon-tin case, the hypotheses on offer are: *architects select for arches and spandrels free-ride* versus *architects select for spandrels and arches free-ride;* and the counterfactuals whose truth or falsity decides the question is: *if there could be arches without spandrels, architects might choose them; but not vice versa.* In the heart case, the hypoth-eses on offer are *Mother Nature selects for pumping and the noises free-ride* versus *Mother Nature selects for noises and the pumping free-rides;* and the counterfactual whose truth or falsity decides the issue is: *if there could have been blood pumps without noises Mother Nature would probably have selected them, but not vice versa.*

Avatars of the 'selection-for' problem continued: 'What is learned?' (the problem of the 'effective stimulus')

We turn to structurally analogous cases that arise in the psychology of learning.

One way to think about learning is that it's a process in which the operation of exogenous forces alter a creature's associations. This is the way that both the British empiricists and American learning theo-rists of the Skinnerian sort actually did think of it (see Chapter 1 for discussion). There is, however, a characteristic difference between the British empiricists' kind of associationism and the learning theorists' kind – a difference much emphasized in the Skinnerian literature. Skinner's kind of learning is the formation of an association between a stimulus and a response, whereas, for the British empiricists, learn-ing is the formation of a relation among mental objects ('ideas'). This raised a problem for learning theorists that the British empiricists hadn't had to face: when you learn an S–R association, *what are the S and the R?* An example should serve to make this problem clear.

Consider what the learning-theoretic literature calls the 'split stimulus' experiment. Get yourself a rat or (more fashionably these days) a pigeon, and a Skinner box. Place the former in the latter. Train the animal on a standard discrimination learning task where SD (the reinforced stimulus; the one in the presence of which the animal's responses are rewarded) is, say, a yellow triangle; and S Delta (the unreinforced stimulus, the one in the presence of which the animal's responses are not rewarded), can be anything you like; perhaps a card with an X on it. When training is completed, the animal produces responses when, and only when, the SD is displayed. (In fact what's actually observed is some asymptotic approximation to that. There is no such thing as a perfect pigeon.) Nothing new so far.

But now, ask yourself: 'what has the animal actually learned when it learns to respond in and only in the presence of the positive stimulus'? Patently there are many possibilities (arguably there are indefinitely many): it may be that it has learned to choose yellow triangles in preference to Xs; or that it has learned to choose triangles in preference to Xs; or that it has learned to choose yellow things in preference to Xs; or that it has learned to choose closed figures in preference to Xs; or that it has learned to choose things with pointy corners in preference to Xs; and so on.

That, then, is how a perfectly typical instance of a discrimination learning experiment raises the question of 'what is learned'. Notice the very close analogy to Gould–Lewontin's question 'When a phenotype changes in consequence of exogenous forces of selection, which (if any) of the new phenotypic traits is an adaptation and which (if any), is a free-rider?'

Experiments that 'split the stimulus' – in effect, applications of J. S. Mill's 'method of differences' (Mill, 1846) – are the canonical way to answer questions about what is learned. For example, such an experiment might be designed to determine whether, after having been trained to discriminate yellow triangles from blue squares the animal responds to ('generalizes' to) blue triangles or to yellow squares, when offered a choice between them. Roughly, if the creature's training generalizes to blue triangles (but not to yellow squares), then what it learned in consequence of training was to respond to triangles; if it generalizes to yellow squares (but not to blue triangles), then what it learned in consequence of training was to respond to yellow things. And so forth. Thus, at least in principle, 'split stimulus' experiments can tell us about what a creature learns when it learns a conditioned response to yellow triangles. That is, indeed, just the kind of experiment that psychologists do use to settle questions about what is learned when such questions arise in practice. It is surely perfectly rational of them to proceed that way.[5]

Our point is not to insist that a creature's history of reinforcement vastly underdetermines what it learned in the course of conditioning (although of course it does).[6] Rather, our point is to emphasize how close is the parallel between, on the one hand, the issues that 'what is learned' questions raise for reinforcement theories and, on the other hand, the issues that free-riding raises for adaptationist theories of evolution. In both cases, the scientist has a choice between hypotheses about which of a number of coextensive traits is selected-for as the result of a certain set of causal contingencies. Is it arches that the architects select for or is it the spandrels? Likewise: is it responses to yellow that conditioning selects for when the SD is a yellow triangle? Or is it responses to triangularity? Or is it both? Or neither? In fact, the 'what is learned' problem *is* a free-rider problem: when a yellow triangle is the SD, does conditioning to yellow free-ride on conditioning to triangularity, or is it the other way around?

Given the structural similarity between the two kinds of free-rider questions, it's hardly surprising that both get answered in the same way, viz. by appeal to relevant counterfactuals. What the adaptationist wants to know is: What would happen if the de facto coextension of arches and spandrels breaks down; which would the architect have selected-for if he had been offered arch-free spandrels on the one hand, or spandrel-free arches on the other? Likewise, what the learning theorist wants to know is: How would the animal have responded if it had been offered a yellow square on the one hand, or a green triangle on the other? The logic of the situation remains the same whether it's an architect or Mother Nature or some psychologist who is doing the selecting; and it's likewise independent of whether it is the organism's phenotype that is being selected-for or its behavioural repertoire. The moral so far is: selection-for problems need to appeal to counterfactuals if they are to distinguish between coextensive hypotheses whether it's the theory of association or the theory of adaptation that raises the question. Likewise in a variety of other kinds of cases that seem, at first glance, quite disparate. We will describe a couple more of them.

Avatars of the 'selection-for' problem continued: 'what is learned?' (the 'effective response')

The moral of the previous discussion was that, just as adaptationist evolutionary theory has a problem about distinguishing the trait selected-for from its free-riders, so learning theory has a problem when it attempts to distinguish the 'effective' stimulus property from its mere correlates. That being so, it's unsurprising that there is a problem of exactly the same sort and magnitude about what the R is in a given instance of S–R conditioning, that is, what a creature has learned to do when it learns a response to a certain stimulus. The classic 'split-response' experiment is due to MacFarlane (1930) and Tolman (1948). A rat is trained on a T-maze on which right turns are rewarded. What does the rat learn when it masters the task? 'Well, it learns to perform a certain behaviour.' Yes, but there are lots of locally coextensive but unequivalent ways to specify the behaviour that it learns to perform. For example, some (rather primitive) versions of learning theory might suggest that what the rat learns to do is make a series of motor gestures (first move the left front leg, then move the right rear leg and so forth). Or it might have learned to turn right; or to turn east; or to turn the same way it turned last time ... or whatever. Which of these is the conditioned response?

Learning theory doesn't say. Learning theory is about how the strength of the association between a stimulus and a response varies a function of reinforcement; but it says nothing about *what the stimulus and the response are.* Accordingly, although you can do experiments that decide among various of the possibilities post hoc, the caveat 'post hoc' is essential, since *the theory doesn't predict the experimental outcomes*. For example, flood the maze with water just enough to require the animal to swim rather than run. What happens isn't intuitively surprising: the rat proceeds, without further training, to swim along the route that it had previously been running through. The converse also applies (running through a route previously learned by swimming). So it can't be that the conditioned response the rat learned in consequence of the training was to *run* to the right. But it's not that the theory of learning predicts the experimental outcome, since it says nothing at all about when (or whether) running and swimming are instances of the same responses. Rather, performing the split-response experiment decides it post hoc. And there are, of course, indefinitely many other possibilities even in this simple instance of reinforcement learning.

Here's a split-response experiment that you can try in the privacy of your own home, with no wet rat required. Put the hand of the Subject (S), palm down, in contact with a device that can deliver a (very, very, very) mild shock. Arrange that a bell goes off a second or so before a shock occurs. S will, we promise you, rapidly learn to move his hand when he hears the bell. But what response (what 'behaviour') did S learn when he learned to do so? To raise his hand? To withdraw his hand? To move his hand in the direction of the ceiling? To move his hand in a direction away from the floor? To move his hand in a direction away from his feet? All of these? None of these?

Notice that these ways of describing what the subject learned are all satisfied by the behaviour that was performed; they are all 'locally coextensive' in the experimental environment. Here, once again, a split-response experiment can decide among alternatives; but here, once again, learning theory per se makes no prediction as to the experiment's outcome. What would happen, for example, if the trained subject, still in contact with the shock device, turns his hand over, palm up? Bell rings. S produces a conditioned response, but what response is that? (As it turns out, while S's fingers flex away from the palm when the hand is right side up, they flex towards it when the hand is upside down.)[7] Another way to put this is that the effect of reinforcement was to vary the strength of the response, that is, reinforcement affects the likelihood that a creature will emit the same behaviour now that it did before. But the question is, *What is to count as* an instance of the same behaviour? This question, like the one about what is to count as a recurrence of the stimulus to which a creature is conditioned, is an abyss that learning theory doesn't even begin know how to bridge. With the response as with the stimulus, the laws of association say nothing at all about what is learned when an S–R association is formed.[8]

So the logic of the 'selection-for' problem recurs: the theory of learning offers two hypotheses, both of which are compatible with the data (in this case, both compatible with the history of S–R conditioning). What decides between them is the counterfactuals about what (would) happen in experiments that split the S or the R.[9]

Avatars of the 'selection-for' problem continued: the naturalization of content

Commonsense explanations of behaviour allow themselves free use of both 'semantic' concepts ('meaning', 'truth', 'reference' and the like), and 'intentional' concepts ('belief', 'desire', 'motive' and the like). But many philosophers and many psychologists view this with deep suspicion. They doubt that such explanations would be countenanced in a 'really first-class' conceptual system. Rather, from the point of view of a developed scientific world view, they would be seen as *façons de parler*, useful in navigating one's day-to-day transactions with the world but not to be taken with full ontological seriousness; or they would be dispensed with entirely, along with witches and phlogiston. Well-brought-up children would learn to say 'it causes my C-fibres to fire' where they now learn to say 'it hurts'. (We are not making this up. See, for instance, Churchland [1981].)

On the other side, there is a cluster of philosophers and psychologists who hold that some intentional explanations are just plain true (and others are just plain false). For them, the problem arises of explaining how to reconcile their literalism about the intentional/ semantic with their overall commitment to some or other version of 'physicalism'; that's to say, to some or other version of the view that the only things that there are in the world are physical things and events. Faced with this dilemma, the research programme for philosophers who are realists about the intentional/semantic is to formulate some condition that is, at least in principle, specifiable in the vocabulary of physics, and that is sufficient for a physical object to be in a semantic/intentional state. For a number of years now, the theory of choice among such philosophers has been some or other version of a 'causal' theory, according to which the content of an intentional state is determined by its causal connections.[10] It was at this point that selection-for problems first showed their head in the argument about whether the intentional/semantic could be naturalized. And, since the examples concerned behaviours that belong to an innate repertoire, the connection to issues about evolution is especially transparent.

Consider the infamous riddle of the frogs and the flies. Frogs snap at flies; having caught one, they then ingest it. It is plausibly in the interest of frogs to do so, since, all else being equal, the overall fitness of a frog that ingests flies is likely to exceed the overall fitness of a frog that doesn't. We suppose it is likewise plausible that frogs snap at flies with the intention of eating them. (If, however, you are unprepared to swallow the attribution of intentions to frogs, please feel free to proceed up the phylogenetic ladder until you find a kind of creature to which such attributions are, in your view, permissible.) Now, intentions are just the sorts of things that have intentional/semantic contents, which serve to distinguish among them. A frog's intention to catch a fly, for example, is an intention *to catch a fly*, and is therefore distinct from, say, the frog's intention to sun itself on a lily pad. Here, then, is a plausible and rudimentary case on which to try out a causal theory of content: perhaps the frog's intention to catch a fly is *about flies* because it is an intention of the kind that is generally caused 'in the right way' (whatever, exactly, that may be) by the proximity of flies.

There are all sorts of things wrong with that suggestion; things that would need to be fixed if a causal account of content is to seem even remotely plausible. Only one of the things that's wrong with it is, however, germane to our present concerns. An intention to catch a fly is not *ipso facto* an intention to catch an ambient black nuisance ('ABN' hereafter); not even on the assumption that these two ways of describing flies are locally coextensive,[11] which, if true, would entail that every snap that's caused by a fly is likewise caused by an ABN, and every snap that's caused by an ABN is likewise caused by a fly.[12] In a nutshell: if the assumption of local coextensivity holds (as, of course, it perfectly well might), then fixing the cause of the frog's snaps doesn't fix the content of its intention in snapping: either an intention to snap at a fly or an intention to snap at an ABN would be compatible with a causal account of what the frog has in mind when it snaps. So causal accounts of content encounter a selection-for problem: If something is a fly if and only if it is an ABN, the frog's behaviour is correctly described either as caused by flies or as caused by ABNs. So, it seems, a causal theory of content cannot distinguish snaps that manifest intentions to catch the one from snaps that manifest intentions to catch the other.

As usual, an appeal to counterfactuals is what breaks the assumed coextension. What would happen in a world where everything is the same as here except that some ABNs aren't flies, or vice versa? Which does the frog snap at in such a counterfactual world? If *those* frogs snap at flies that aren't ABNs, then (all else being equal) our frogs must be fly-snappers; if those frogs snap at ABNs that aren't flies, then (all else being equal) our frogs must be ABN-snappers. The moral is: the cause doesn't determine the content; but maybe the cause *together with the relevant counterfactual* does.

A point we want to emphasize is that this selection-for problem about content implies a corresponding selection-for problem about natural selection. The frog's disposition to fly-snap belongs, after all, to its behavioural phenotype; and the evolution of phenotypes, behavioural or otherwise, is what the theory of natural selection is committed to explaining. So, then, which did the frog evolve: a disposition to snap at flies or a disposition to snap at ABNs? (Or both? Or neither?)

Should we say that the selection of fly-snapping free-rides on the evolution of ABN-snapping, or was it the other way around? And what was it about the selection pressures on frogs that determined whether they evolved fly-snapping phenotypes (if that's what they did) or ABN-snapping phenotypes (if that's what they did)? If, in short, there are phenotypic traits that are distinguished by their content, then for each such trait there must be a corresponding distinction between evolutionary histories. What do such distinctions consist of? What makes the relevant counterfactuals true and what makes them false?[13]

Interim summary and prospectus

Once you've noticed that there are selection-for problems, you start to see them everywhere. We've chosen examples from what are often treated as quite different fields of inquiry, and there are many others we might have chosen instead. In all such cases, the same logic applies: there are competing explanations of why Xs are P: *Xs are P because they have property F* and *Xs are P because they have property G*. By assumption, exactly one of these explanations is true; and likewise, by assumption, properties F and G are (locally) coextensive. In effect, the situation is either that *their being F explains Xs being P and their being G free-rides on their being F*, or it's that *their being G explains Xs being P and their being F free-rides on their being G*. Because F and G are, by assumption, coextensive, facts about the *actual* world don't choose between the explanations; but certain counterfactuals do. If *Xs that are F are P* is true in a possible world where not all Xs that are F are G, then (all else being equal) it is Xs being F (rather than Xs being G) that explains Xs being P in the actual world. And vice versa, as usual.

We can now announce our overall polemical strategy: we started this chapter by recalling Gould and Lewontin's insight that a theory of natural selection must somehow allow for the possibility of phenotypic traits that are not adaptations. We think that Gould and Lewontin were entirely right about that, but we think they missed a deeper point: once the character of selection-for problems is properly understood, it becomes apparent that the question that phenotypic free-riding raises cannot be answered within the framework of adaptationist theories of evolution. If that's right, then adaptationism simply cannot do what an evolutionary theory is supposed to do: explain how phenotypic traits are distributed in populations of organisms. Equivalently: the theory of natural selection cannot predict/explain what traits the creatures in a population are selected-for.[14]

The argument unfolds in three stages. First we suggest a diagnosis: we want to make clear what it is that causes a theory to give rise to selection-for problems. If we've got the diagnosis right, it leads quite directly to a way to distinguish the kinds of selection-for problems that have solutions from the kinds of selection-for problems that do not. Finally, we'll argue that (on quite plausible empirical assumptions) the selection-for problems that afflict adaptationism are of the unsolvable kind. Roughly, the rest of this chapter is directed to setting out the first two phases of the argument. The third phase will occupy Chapter 7.

Where do 'selection-for' problems come from?

The question is rhetorical; we think that reflection on the sorts of cases we've been describing suffices to make the answer clear: selection-for problems (hereafter referred to as SFPs) turn up when explanations (or theories, or whatever) require distinguishing between the causal roles[15] of coextensive properties.[16] For example: What was it about the teleology example that gave rise to its SFP? The answer, according to us, turned on the coextension of the properties *being a blood pump* and *being a heart-noise-maker*. Outside the laboratory and the emergency room, anything that has either property has the other. Accordingly, the SFP that arose: Which property is the *function* of the heart; and what is it about the example that makes it *that* property rather than the other? The same considerations arose with respect to the distinction between properties of a domed church that are selected-for and properties that merely free-ride. So, in the Gould–Lewontin example that we started with, *having spandrels* and *having arches* are coextensive in certain kinds of churches; in that kind of church, there are spandrels if and only if there are arches. Still the intuition is very strong that it's the arches that explain the spandrels, not the other way around; and that it is the arches not the spandrels that the architect selected for when he designed the church. The SFP that the example raises is: What is it about arches or spandrels (or architects) that supports these intuitions? We leave it to the reader to review the other cases of SFPs we have discussed and see how well the diagnosis fits.

Yes, but what does all this stuff have to do with whether or not natural selection is the mechanism of the evolution of phenotypes?

Fair question. The first point to notice is that the problem of distinguishing coextensive traits arises very urgently whenever a selectionist theory wants to explain *which of its phenotypic traits makes a certain kind of creature fit in a certain kind of ecology.* Such explanations are, of course, at the core of selectionist theories: creatures survive and flourish because, in the creature's ecology, certain of its phenotypic traits are 'correlated with fitness'. The scientific problem is to figure out which traits these are, and why they affect fitness in the ways they do. None of these remarks are tendentious as far as we know.[17]

But, of course, it can happen – and it very often does – that phenotypic traits that affect fitness are confounded with phenotypic traits that don't. In such cases, *both* of the traits are 'correlated with fitness'; trait T is, as it were, *directly* correlated with fitness, and trait T′ is correlated with fitness indirectly by virtue of its correlation with T. That's hardly surprising; in fact it's just a way of describing cases in which T is selected-for and the selection of T′ free-rides. Very well; but now we are faced with the question: *How could* a process of selection result in such a situation? After all, the adaptationist story was supposed to be that traits are selected-for when they are correlated with fitness, which by assumption *both* T and T′ are. So how could it be true that either is selected-for and the other is not? In particular how could it be true that one of the traits is free-riding on selection for the other?

'But didn't you just say that T and T′ affect fitness in different ways: one affects fitness directly and the other does so via the first? So maybe we just have to amend the adaptationist principle; we have to say "phenotypic traits are selected for their *direct* effects on fitness." Won't that do just as well?'

Well, yes, we did say that; but we didn't understand it when we said it and we still don't. For what, exactly, is this 'directness' such that selection-for can act on one but not the other of two coextensive traits because one has a connection to fitness that is direct but the other doesn't? One might suppose that this is where the counterfactuals come in. Suppose T and T′ are coextensive. Then if T is selected 'directly' and T′ is selected by virtue of its correlation with T, then the following counterfactual ought to be true: *if T hadn't been selected, T′ wouldn't have been selected either* (*but not vice versa*). In effect, the idea is that T′ is correlated with fitness via a chain of causes and effects that runs through T and ends at some modulation of fitness. If T were removed, then the effect of T′ on fitness would vanish; but if T′ were removed, then (all else being equal) T would remain intact and its effect on fitness would be unaltered. Of course, neither T nor T′ actually *is* removed; by assumption, the two are coextensive in the actual world and their coextension is broken only in worlds that are counterfactual. Still, we can now see how selection could distinguish between T and T′ *so long as selection is sensitive to the counterfactuals about whether removing one of them affects the adaptivity of the other.*

But we are still in the woods. For the question now arises: *How could* selection be sensitive to the consequences of *counterfactually* removing T but not T′ (and/or the consequences of counterfactually removing T′ but not T) if, in point of fact, neither T nor T′ actually *is* removed? The answer is that it couldn't. Selection cannot, as a matter of principle, be contingent upon (merely) counterfactual outcomes. That, in a nutshell, is why we think that selectionism cannot be true.

We want to remind you of a point we remarked on in Chapter 1:

selection is a *local* process; only ecological variables with which it causally interacts can exert selection pressures on a population. So, for example, future events cannot (unless actual events foreshadow them); and past events cannot (unless traces of them persist in the present); and events from which the population is geographically isolated cannot;[18] and so forth. The aspect of locality that matters for our present purposes, however, is that the outcomes of merely *counterfactual* events cannot exert selection pressures: merely possible predators do not affect the evolution of a population (although, *actual* predators are quite likely to do so). The number of rabbits in Australia is unaffected by the number of foxes in England. That's because the predations of the one on the other are all merely counterfactual, and possible-but-non-actual events do not exert selection pressures. The situation would, we suppose, be quite different if the two populations came into *actual* causal contact. All else being equal, there would soon be more foxes and fewer rabbits; and, quite likely, both would soon run faster than they did before.

Although the insensitivity of the course of evolution to merely counterfactual goings-on is sufficiently obvious, its implications for raising selection-for problems in evolutionary theory has gone widely unappreciated. We've been arguing that the distinction between traits that are selected-for and coextensive traits that free-ride upon them turns up in the relevant counterfactuals; white polar bears would catch fewer fish if their environment turned green, so there's a prima facie case that white polar bears were selected-for matching their environments and not for their colour; and so forth. We may now add that the same point holds for any trait that is even *locally* coextensive with a trait that's selected for: unless it affects the actual causal interactions between a population of creatures and its ecology, it cannot affect the evolution of that population. Putting all this together, we get the following:

(1) Selection-for is a causal process.
(2) *Actual* causal relations aren't sensitive to *counterfactual* states of affairs: if it *wasn't* the case that A, then the fact that it's being A

would have caused its being B doesn't explain its being the case that B.[19]

(3) But the distinction between traits that are selected-for and their free-riders turns on the truth (or falsity) of relevant counterfactuals.

(4) So if T and T′ are coextensive, selection cannot distinguish the case in which T free-rides on T′ from the case in which T′ free-rides on T.

(5) So the claim that selection is the mechanism of evolution cannot be true.

'Where did you get (5)?' you may wish to ask. 'Why is it so important that a theory of evolution should reconstruct the distinctions between free-riders and hangers-on?' We've already seen the answer: evolutionary theory purports to account for the distribution of phenotypic traits in populations of organisms; and the explanation is supposed to depend on the connection between phenotypic traits and the fitness of the creatures whose phenotypes they belong to. But, as it turns out, when phenotypic traits are (locally or otherwise) coextensive, selection theory cannot distinguish the trait upon which fitness is contingent from the trait that has no effect on fitness (and is merely a free-rider). Advertising to the contrary notwithstanding, natural selection *can't be* a general mechanism that connects phenotypic variation with variation in fitness. So natural selection can't be the mechanism of evolution.

If that's not bad enough for your taste, please do wait for Chapter 7: things will get worse. Much worse. Meanwhile, we want to remind you of one last point about the character of selection-for problems. Consider, once again, the Gould–Lewontin paper about arches and spandrels. We have simply assumed, as they do, that it's the arches rather than the spandrels that are selected for, and that that's because it was the arches and not the spandrels that the architect had in mind when he designed the building. So it looks as though we must be saying that, although evolution is insensitive to the distinction between traits selected-for and their free-riders, architects aren't. If that's right, then architects are an exception to an otherwise very plausible principle that controls causal explanations in general and to which evolutionary explanations must therefore conform, viz. that what happens in the actual world is unaffected by the truth (or falsity) of (mere) counterfactuals. Hasn't something gone wrong?

That is indeed what we're saying; and nothing has gone wrong. The salient difference between architects and the processes of evolutionary selection is that *architects have minds and evolutionary processes do not.* Minds are useful things to have; it's among their virtues that they can represent things that *didn't* happen; or things that happened *a long time ago*; or things that happened *far, far away*; or things that *will* happen; or things that *might* happen; or that would happen *if* ..., etc. This includes, of course, counterfactual events and their counterfactual effects. So, as previously remarked, an architect can say to himself: 'If I were to take the spandrels out, I would have to take the arches out too; and if I were to take the arches out, the dome would fall down. Given a choice between not removing the spandrels and having the dome fall down, I opt for the former.' This is a classic case of thinking things through before deciding how to act. It's why our thoughts can 'die in our stead'.

We don't claim to know *how* minds go about representing counterfactual events (or future or past events either); or, for that matter, how they go about representing things that are right in front of their noses. Suffice to say that they can and do. It's therefore unsurprising that, when Gould and Lewontin wanted a good, firm, entirely intuitive example of the selected-for/free-rider distinction, they chose a case of *mental causation,* a case in which there *actually is* an 'intelligent designer'. So, too, did Darwin when he came to explaining how natural selection works; the idea was that natural selection works just like breeding, *except that, in the case of natural selection, there isn't any breeder.*[20]

But there isn't, of course, an intelligent designer in the case of evolution; indeed, there isn't any designer at all. That makes the situation very tricky for the theorist. 'Selection-for a trait' is the pivotal notion in adaptationism; the central idea of adaptationism is that it provides for an entirely *naturalistic* mechanism of selection for phenotypic traits. It is thus of prime importance that the success of the project of explaining phenotypic traits in terms of what they are selected-for *doesn't depend on assuming that selection-for a trait is the effect of mental causes.* Selection-for is, of course, the effect of mental causes in the case of the architect; and it is again the effect of mental causes in the case of the breeder. Darwin (and, we suppose, Gould and Lewontin) thought that he could *start with mental processes and then get to natural selection by abstracting the minds away.* But that, in a nutshell, is what we are saying can't be done. For, a theory of evolution must be able to distinguish the causal powers of coextensive traits; and (as far as we know) the causal powers of coextensive traits can be distinguished only by appealing to distinctions among counterfactuals; and (as far as we know) only minds are sensitive to distinctions among counterfactuals. We now add that an essentially identical line of argument works for *any* evolutionary explanation that (implicitly or otherwise) invokes the notion of selection for a trait. The distinction between locally coextensive traits *can* be drawn if counterfactuals are taken into account. But counterfactuals have their effects on happenings in the actual world only via the mediation of minds, and its common ground that no minds mediate the processes of natural selection. It is very hard indeed to get an account of evolution that actually does get the *deus* out of the *machina*. Even Darwin didn't know how to do it. And nor, of course, do we, if we have to stick to adaptationism. So we've given up on adaptationism.

So there would seem to be a problem. Is there no way out of it? We think that the prospects of finding one aren't good; we think this pill will just have to be swallowed. Part of Chapter 7 is about what's wrong with some standard adaptationist proposals to the contrary.

7

NO EXIT? SOME RESPONSES TO THE PROBLEM OF 'SELECTION-FOR'

Chapter 5 took 'population thinking' for granted: evolutionary theory aims to express the generalizations according to which phenotypic traits vary lawfully as a function of ecology. We needed this assumption for the main line of argument, which was that getting it right about which empirical theories are true means getting it right about which counterfactuals are true; and that selection theory is intrinsically unable to distinguish true counterfactuals from false ones in cases that are relevant to the individuation of phenotypic traits.

But we can (just barely) imagine someone replying like this: 'Bother generalizations and double-bother hypotheticals with false antecedents. What I care about getting right are the actual facts about the actual world; and I don't care about anything else'. 'If it's true', this hypothetical person continues, 'that other theorists have more grandiose ambitions, perhaps that's because they hang out with philosophers. So much the worse for them.'

What ought one to say to such a person? We think it's that you can't get it right about the actual world unless you get it right about counterfactual worlds; not, anyhow, so long as you are running a theory about traits; which is, indeed, just what Darwinists claim to be doing. We think this is important, so before we continue, we'll provide a recapitulation in brief of the argument from Chapter 5 that it depends on. It's a short argument.

Recapitulation in brief

(1) Since selection is a local process, it follows that if traits are locally coextensive (a fortiori, if they are coextensive *tout court*), they must have the identical correlation with fitness. In particular, they must be correlated with the same outcomes of all *actual* competitions. In the actual ecology, frogs that have the phenotypic trait of *being fly-snappers* won't win competitions with frogs that have the phenotypic trait of *being ambient black nuisance-snappers*; polar bears that are selected-for being white won't win competitions with polar bears that are selected-for matching their environments, etc. However, fly-snappers do win competitions with ambient black nuisance-snappers in possible but non-actual worlds where the ambient black nuisances aren't flies, and polar bears that are selected for matching their environment do win competitions with white polar bears in non-actual worlds where the environment is green. In effect, when evolutionary accounts are coextensive in their application to actual outcomes, one distinguishes between them by reference to their application in counterfactual outcomes. This, as we remarked in Chapter 1, is the logic of appeals to the 'method of differences' in deciding between competing theories. Working scientists exploit this tactic all the time.

(2) So the issue is whether a theory of evolution by natural selection can predict the outcomes of merely counterfactual competitions. If it can't, it won't be able to decide, as between locally coextensive phenotypic traits, which of them explains effects on fitness.

Moral: a theory that doesn't determine the truth values of relevant counterfactuals cannot explain the distribution of traits in the actual world.[1]

That being so, the question arises: What, according to adaptationist accounts, supports the relevant counterfactuals about the evolution of phenotypic traits? What, according to adaptationists, does the truth or falsity of these counterfactuals consist of? Since this question is remarkably under-discussed in the literature (Sober [1993] is an honourable exception; see below), we're more or less on our own. We can think of four suggestions one might try – all of which are pretty clearly unsatisfactory, and one of which is flat-out preposterous. It is, of course, possible that someone will think of a fifth (or a sixth, or a seventeenth). But don't hold your breath.

First option: give Mother Nature a chance

There's a sort of analogy between what natural selection does when it culls a population and what breeders do when they select from a population those members that they encourage to reproduce. This analogy was emphasized by Darwin himself, and it has been influential in the popular sort of adaptationist literature ever since. Suppose Granny breeds zinnias with the intention of selling them on Market Day. Then Granny is selecting zinnias for their value on the market, and not, say, for the elaboration of their root systems. This is so even if, as a matter of fact, it's precisely the zinnias with elaborate root systems that sell at the best prices. Likewise, the fact about her intentional psychology that explains which zinnias Granny chooses when she sorts them is that she is interested in selling them, and not that she is interested in their having lots of roots. (Granny may not even know about the connection between market values and root systems. Or, if she knows, she may not care.[2]) In short, since Granny is in it for the money and not for the roots, there is a matter of fact about what traits she selects for when she selects some of the zinnias and rejects others. What Granny *selects-for* is: *whatever it is that she has in mind when she does her selecting*.[3]

So, then, perhaps we should take the analogy between natural selection and selective breeding at face value. Perhaps we should say of natural selection just what we said of Granny: that what it selects for is whatever it has in mind in selecting. Notice that the counter-factuals fall out accordingly: if Granny is interested in high market value rather than big roots, that decides what she *would* do (all else being equal) in a world in which the saleable zinnias are the ones with short roots, or no roots, or green roots with yellow spots, or whatever. Likewise, if natural selection has it in mind that there should be lots of frogs that catch flies, then, in the actual world where the flies or ambient black nuisances (ABNs) are mostly flies, it favours both frogs that snap at flies and frogs that snap at ABNs. But in the counterfactual world where the flies-or-ABNs are mostly ABNs, natural selection will favour only the frogs that snap at flies.[4] So, then, perhaps we should think of natural selection as Granny writ large, and say of the one what we said of the other: what natural selection selects-for is whatever it has in mind in selecting.

That, at least as much as stuff about designs needing designers, is the thought that explains the prominence of anthropomorphized avatars of natural selection in the adaptationist literature: Mother Nature, the Blind Watchmaker, the Selfish Gene or, for that matter, God.[5] All of these are supposed to be 'intentional systems': that is, they are supposed to have intentions in light of which they act.[6] So, to construe natural selection on the model of artificial selection is to make room for a distinction between selection having it in mind to propagate frogs that snap at flies and selection having it in mind to propagate frogs that snap at flies-or-ABNs; precisely the distinction that we need to make room for if we are going to make sense of traits being selected-for.

When it's put that baldly, however, it is perfectly obvious what's wrong with this line of thought: natural selection doesn't have a mind. A fortiori, it has nothing in mind when it selects among frogs.[7] Likewise, if genes were intentional systems, there would be an answer to the question of whether natural selection favours creatures that really do care about the flourishing of their children or creatures that really care only for the propagation of their genotypes. All you would have to do, if you wanted to know which sort of creatures we are, would be to find out which of these phenotypes our genes prefer.

If genes were intentional systems, or if there were a Mother Nature who selects with ends in view, then there would be a matter of fact about which traits they select for and which traits are merely coextensive with the ones they select for. That's the good news. The bad news is that, unlike natural selection, Mother Nature is a fiction; and fictions can't select things, however hard they try. Nothing cramps one's causal powers like not existing. Likewise, *mutatis mutandis*, the genes that make you try to cause your children to flourish (if, indeed, there are such genes) couldn't care less about why you want your children to do so. They couldn't care less about that because they don't care at all about anything.

We want to make clear just what we're claiming is the connection between, on the one hand, construing natural selection on the model of intentional selection and, on the other, making sense of natural *selection-for*. The point here is not (did we say that loud enough?) *IS NOT* that there is no design without a designer (although indeed there isn't; see Chapter 6 note 17). Rather it's that the individuation of traits depends on the truth of counterfactuals: since (by assumption) every fly-snap in the actual world is an ABN-snap and vice versa, selection between fly-snappers and ABN-snappers must be sensitive to the counterfactual consideration that ABN-snapping gathers no flies in worlds where the ABNs are BBs, rather than flies. It's a nice thing about intentional systems that they *are* sensitive to *merely counterfactual* contingencies. It means that beliefs can take account of what the outcomes of actions *would be if* ... and the believer can then act accordingly.[8] So, thinking of selection as an intentional process is one way to bring into play the counterfactuals that we need to make the distinctions that we need in order to individuate phenotypic traits. (There are other ways; we'll get to that presently.)

To repeat: the advantage of leaning on Mother Nature is not that she's complex enough, or intelligent enough, or conscious enough, to be in the trait-selection game; it's that (by assumption) she's an intentional system, and intentional systems are sensitive to counterfactual

WHAT DARWIN GOT WRONG

outcomes. All that being so, it would be a great help to adaptationists if there were a Mother Nature. However, since there isn't one, she is a frail reed for them to lean on. Ditto the Tooth Fairy; ditto the Great Pumpkin; ditto God.[9] Only agents have minds, and only agents act out of their intentions, and natural selection isn't an agent.

You may think the preceding speaks without charity; that we are, in fact, shooting in a barrel that contains no fish. Surely, you may say, nobody could really hold that genes are literally concerned to replicate themselves? Or that natural selection literally has goals in mind when it selects as it does? Or that it's literally run by an intentional system? Maybe.[10] But, before you deny that anybody could claim any of that, please do have an unprejudiced read through the recent adaptationist literature[11] (especially in evolutionary psychology). Meanwhile, we propose to consider a different way of arguing that adaptationism can ground the counterfactual outcomes that distinguish fly-snapping frogs from ABN-snapping frogs, thus providing a paradigm for selectionist accounts of the content (and the teleology) of mental states.

Second option: laws of selection

Laws can support counterfactuals. That's most of what they do for a living; arguably, it's what makes them different from mere true empirical generalizations. So, then, suppose there is a law that says that (in such and such circumstances) t_1s are selected in competitions with t_2s. If that's a law, then (tautologically) it holds in all nomologically possible states of affairs; which is to say that it determines the outcome of any nomologically possible t_1 versus t_2 competition, including ones that are merely counterfactual. None of that should seem surprising. Or at least none of it should on the assumptions that laws are relations among properties and that properties that aren't instantiated in the actual world may nevertheless be instantiated in some nomologically possible other world. So laws of selection might support the counterfactuals that are required to vindicate the distinction between selection and selection-for. So all may yet be well. So the story goes.

Presumably, the paradigm of the kind of law we're looking for would be something like: 'All else being equal, the probability that a t_1 wins a competition with a t_2 in ecological situation E is p.' But are there such laws? We doubt that there are (although, it is, of course, in large part an empirical issue);[12] a priori argument won't decide it one way or the other. But we can think of several reasons why there might seem to be laws of selection even if, as a matter of fact, there are none. We want to look at some of these, because, unlike the idea that natural selection is an intentional system, the suggestion that counterfactuals about selection are grounded in laws of selection isn't nutty; it's just (according to us) untrue.

The fallacy of the Swiss apple

It's a thing about laws that they aspire to generality: in the paradigm cases, a law about Fs is supposed to apply to instances of F as such. Conversely, to the extent that a generalization applies not to Fs as such but only to Fs in such-and-such circumstances, it's correspondingly unlikely that the generalization is a law (or, if it is a law, it's correspondingly unlikely that it's a law about Fs as such). We take that to be common ground. But if it's right, then quite likely there aren't any laws of selection. That's because who wins a t_1 versus t_2 competition is massively context sensitive. (Equivalently, it's massively context sensitive whether a certain phenotypic trait is conducive to a creature's fitness.) There are a number of respects in which this is true – some obvious, some less so.

For example, it's obvious that no trait could be adaptive for creatures across the board. Rather, the adaptivity of a trait depends on, among other things, the ecology in which its bearer is embedded. In principle, if a trait is maladaptive in a certain context, *you can fix that either by changing the trait or by changing the context.*[13] Is being green good for a creature's fitness? That depends on whether the creature's background is green too. Is being the same colour as its background good for a creature's fitness? That depends on whether the camouflage that makes it hard for predators to find also makes it hard for the creature to find a mate.[14] Is it good for a creature's fitness to be big? Well, being big can make it hard to flee from predators. Is it good for a creature to be small? Perhaps not if its predators are big. Is it good for a creature to be smart? Ask Hamlet. (And bear in mind that when selection has finally finished doing its thing, it is more than likely that the cockroach will inherit the Earth.) Whether a trait militates for a creature's fitness is the same question as whether there's an 'ecological niche' for creatures that have the trait to occupy; and that *always* depends on what else is going on in the neighbourhood. Is it good to be a square peg? Not if the local holes are mostly round.

We want to emphasize that our point isn't just that if there are laws about which traits win which competitions,[15] they must be 'all else being equal' (*ceteris paribus*) laws. To the contrary, we take it to be true quite generally that laws of the non-basic sciences hold only 'all else being equal'. If that's so, it's not a complaint against the putative laws of selection that they do too.

We think, however, that the present considerations go much deeper. Perhaps, in the circumstances, a little philosophy of science may be permissible.

To a first approximation, the claim that, 'all else being equal, Fs cause Gs' says something like: 'given independently justified idealizations, Fs cause Gs reliably.'[16] The intuition in such cases is that, underlying the observed variance, there is a bona fide, reliable, counterfactual-supporting relation between *being F* and *causing Gs*, the operation of which is often obscured by the effects of unsystematic, interacting variables. The underlying generalization comes into view when the appropriate idealizations are enforced (typically in the experimental laboratory). By contrast (so we claim) there just aren't any nomological generalizations about which traits win competitions with which others. It simply isn't true, for example, that being big is in general better for fitness than being small (except when there are effects of interacting variables); or that flying slow and high is in general better for fitness than flying fast and low (except when there are effects of interacting variables); or that being monogamous is in general better for fitness than being polygamous (except when there are effects of interacting variables), etc. It's not that the underlying generalizations are there but imperceptible in the ambient noise. It's rather that there's just nothing to choose between (for example) the generalization that being small is better for fitness than being big and the generalization that being big is better for fitness than being small. Witness the fact that the world contains vastly many creatures of both kinds.[17] We don't doubt that there are explanations of why competitions between creatures with different traits come out the way they do; but such explanations don't work by subsuming the facts they explain under general laws about the relative fitness of the traits.[18] (We'll say something further on about how we think they actually do work.)

Nor is that by any means the whole story about the context dependence of being a trait that's selected for. For one thing, traits in isolation don't get selected for at all; that is next door to a truism. The truism it's next door to is that creatures don't have traits in isolation; what they have is *whole phenotypes*, and, quite possibly, whether a trait is fitness enhancing depends a lot on what phenotype it's embedded in. That too is practically a truism; but it's one that game-theoretic models of evolution (for example) have a bad habit of ignoring.

'What would happen if a population of ts were to invade a populations of not-ts?' That depends a lot on what other differences there are between the ts and the not-ts. 'Yes, but *all else being equal,* what would happen if a population of ts were to invade a population of not-ts'?[19] Who knows? In any case, since all else practically never is equal, the question is likely to be academic. Indeed, it may quite well be more or less academic *as a matter of natural law.* Suppose it's nomologically necessary that t phenotypes include property p and that not-t phenotypes don't. Then there *can't be* a competition between ts and not-ts where all else is equal. Since it's presumably a law that there are no flying pigs, who cares what would happen if pigs could fly?

A way of putting this point is that, when you're thinking about the likelihood that there are laws about which phenotypes win which competitions, it's important to bear in mind that outcomes of competitions are interaction effects. It can't be assumed, except in aid of the most austere idealization, that phenotypic traits are in general mutually independent. Phenotypes aren't like Swiss apples; they don't, when you tap them, fall apart into discrete constituents. Evolutionary processes can select for a certain phenotypic trait only insofar as its interactions with other phenotypic traits can be discounted. How far is that? Once again, nobody knows, but it's surely nothing like the general case.

A helpful way to see how massively context dependent generalizations about the fitness of phenotypic traits are is to consider a point we remarked on previously very often: the values of a trait parameter that are viable for one kind of creature can differ radically from the values that are viable for another. Consider size: size affects fitness; but it doesn't follow that there are laws that determine the fitness of a creature as a function of its size. To the contrary, a vast variety of sizes are viable depending not just on what ecology they're in but also on what *other phenotypic traits they have*. The smallest animal[20] (a kind of insect) is said to be about 1.7 millimetres long. The largest animal (the blue whale; probably the largest animal that there has ever been) is 80–90 feet long. Intermediate creatures are scattered very widely throughout the size continuum, which is to say that a size that is viable for one kind of creature need not be viable for creatures of other kinds. (If you're going to weigh several tons, you'd better have a lot of skeleton, or live in the water, or both.) That's to say that the effect of size on fitness must depend on its interactions with other phenotypic variables. Moreover, since a kind of creature of a certain size may be viable in one ecology but not in another, it follows that the interaction of size with other phenotypic variables itself interacts with ecological variables to determine fitness. How many such interactions might there be?

Phenotypes aren't *bundles* of traits; they're more like *fusions* of traits (see Chapter 6). Prima facie, the units of phenotypic change are whole phenotypes. The same considerations hold, of course, in respect of the ecological variables with which phenotypic variables interact. They, too, come in fusions, not bundles; if you tinker with one it's anybody's guess what others you may have to change too (cf.

our lamentable and frightening inability to predict the likely consequences for the viability of marine species of a few degrees of change
in the temperature of the oceans).

In short, the size of a creature quite likely affects its viability. It
doesn't follow that there's a law that determines a creature's fitness
as a function of size. It doesn't even follow that there's a law about
the interaction between, on the one hand, a creature's size and, on
the other, its fitness in a specified ecology; in fact, it's hard to see how
there could be such a law, since animals of all sorts of sizes are often
viable in the same ecology.

To be sure, none of that actually *shows* that there aren't laws of
selection: there may be, on the one hand, units of phenotypic change
and, on the other hand, units of ecological change; and there may
be laws that connect the two. But there's no reason to suppose, as
adaptationists routinely do, that the units of phenotypic change are
anything like what we generally think of as individual phenotypic
traits; and there's no reason to suppose that the units of ecological
change are anything like what we generally think of as ecological 'features'. This matters a lot, since it means that, even if there are laws
of selection, we can't take for granted that they support the sorts
of counterfactuals that a theory about the individuation of traits in
populations requires; that is, counterfactuals about what happens to
fitness in nearby worlds where a given phenotypic trait (or ecological
feature) is altered and everything else is left intact. If, as we suppose,
it is often a matter of empirical law that phenotypic (or ecological)
traits are coextensive, then there may be *no* nearby worlds in which
one such trait is altered and everything else is left intact.[21]

Third option: Sober's sieve

We got into all this because if you have a theory about traits you will
need to appeal to counterfactuals to individuate them, and the only
way to support the counterfactuals that Darwin's theory needs would
be to assume either that natural selection is an intentional system
(which is, at best, preposterous) or that there are laws of selection

(which is, at best, implausible). But it has several times been suggested to us that an example of Elliot Sober's (Sober, 1993, pp. 98–100) provides a third option. This requires some discussion.

Roughly, Sober imagines that a mixed batch of marbles that differ in size and colour is put through a sieve, the holes of which are no larger than the smallest marble.[22] Suppose that all and only the small marbles are red and all the others are some different colour. Then all and only the red marbles will pass through the sieve, even though, as Sober points out, there is a strong intuition that his device sorts not for colour but for size. In effect, the example purports to illustrate the *select/select-for* distinction in miniature, and to do so in a way that vindicates the existence of the distinction. What it *sorts* are the marbles; what it sorts them *for* is their size. The problem is to figure out what grounds these intuitions.

Actually, we don't think that's awfully hard: we know what Sober's sieve is sorting for because *we know how it works;* that is, we know the relevant fact about its endogenous structure. In particular, we know that what it does to the marbles is independent of their colour but not of their size.[23] By contrast, the laws of evolution that adaptationism requires are supposed to express generalizations about which *ecological* variables determine the relative fitness of phenotypes. The idea is that it's ecological laws – laws that apply by virtue of a creature's *exogenous* relations – that support counterfactuals about which traits the creature *would* be selected for *if* it had them. And ecological laws tell us nothing at all about endogenous features (except that they generate phenotypic variations at random).

In short, at best the intuitions about Sober's sieve show us how to draw the select/select-for distinction when *the mechanism that mediates the selection is specified.* That being so, they tell us nothing about how to draw it within the framework of adaptationist assumptions; that is, where we know which phenotypic traits covary with fitness but we (typically) don't know what causal mechanism mediates the covariance. The problem of free-riding *just is* that in some cases phenotypic traits covary with changes of fitness because they cause them, but in other cases they covary with changes of fitness only because they covary with whatever it is that causes them. In Sober's example we're *given* the mechanism that connects phenotypic properties with the outcomes of sorting; but in typical adaptationist explanations, we aren't. So appeal to Sober's sieve *doesn't* show that adaptationist explanations can reconstruct the distinction between selection and selection-for.

Sober's machine doesn't reconstruct the select/select-for distinction even if we grant the intuitions it elicits. But as a matter of fact, we shouldn't grant them because the intuitions that Sober appeals to are themselves entirely illusory; if you share them, that's because (as Wittgenstein might have said) you have a picture in your head. To see that, ask yourself: which of the marbles in Sober's example correspond to the 'fit' creatures in selectionist accounts, and which of the marbles are the ones selected against? It's clear that Sober has in mind that the marbles that are selected are the ones that can get through the holes and reach the bottom, whereas the marbles that are selected against are the ones that the sorting leaves on the top. But notice that this way of describing what happens is entirely arbitrary *even if the mechanism that performs the sorting is exhaustively specified*. Sober must be thinking of the kind of sorting that goes on when you sift flour: the mixed stuff goes into the top of the machine, the good stuff comes out of the bottom of the machine; and what's left behind is the bad stuff. Suppose, however, one thinks not of sifting flour but (for example) of panning for gold. In that case, it's the bad stuff that goes to the bottom and the good stuff that is left as the residuum. *What the machine is sorting-for depends on what the prospector had in mind when he did the sorting.* This is exactly what one ought to expect; prospectors are intentional systems too.

In short, Sober's sieve suffers from the indeterminacy that you always get when you try to interpret an intentional process in a domain that is specified extensionally. Say, if you like, that the machine sorts for size rather than for colour. But, since all and only red marbles stay on top, you might equally say that the machine is sorting for colour rather than size. In the machine Sober describes, sorting for size *isn't distinguishable from* sorting for colour. That's because sieves, unlike prospectors, *aren't* intensional systems. It's the usual thing: a characterization of the extension of a sort simply doesn't determine what it's a sort-for; extensions don't determine intensions, to put it in the philosophers' jargon. Sober's sieve isn't exempt from that hard truth (and nor is Darwin's).[24]

This all makes a point that we think it is well to keep in mind. It's one thing to claim that there are laws that determine the course of evolution; it's something quite else, and quite a lot stronger, to suppose that they are (what we've been calling) 'laws of selection'. That evolutionary processes are subsumed by *some laws or other* follows simply from the assumption that physics is true of everything; if it is, then the phenomena of evolution fall under the laws of physics, along with everything else; that ought to be common ground in the present discussion.

Physicalism per se can't vindicate adaptationism (nor, by the way, can it vindicate learning theory; as usual, the Skinner/Darwin analogy is exact). Arguably, what determines which trait was selected-for is which laws governed the selection: given the laws, the counterfactuals follow; given the counterfactuals, you can distinguish a trait that's selected-for from a trait that isn't. But physicalism isn't committed to any particular inventory of laws; it says only that every causal interaction falls under *some physical law or other*. It follows from physicalism that *if* there is such a process as natural selection, it falls under physical laws (inter alia). But that says *nothing at all* about whether there is such a process. So the next time someone tells you that adaptationism must be true because it is required by the 'scientific world view', we recommend that you bite his or her ankle. We return to the main line of work.

Where we've got to so far: 'sorting-for' is an intensional process. If there is an agent doing the sorting-for, that would account for its intensionality; but, in the case of evolutionary adaptation, there of course isn't an agent. Alternatively, if there are laws of adaptation (laws about the relative fitness of phenotypes), that too would account for the intensionality of sorting-for.[25] But it looks like there aren't any.

Fourth option: maybe adaptationism is not a theory after all; maybe it is just a theory schema

How about treating the theory of natural selection as a theory schema, perhaps along the following lines: adaptationism makes the empirical claim that, for each phenotypic trait (or, for each phenotypic trait that is an adaptation) there is an ecological problem of which the trait selected-for was the solution. Adaptationism per se does not say, in any particular case, either which phenotypic trait was selected-for or which problem it was selected-for solving. But it does say that, in any bona fide case of adaptation, there always is such a trait and such a problem. This claim constitutes the basic empirical commitment of the theory.

We think that's fine if, *but only if*, 'adaptation', 'selection-for', etc., are independently defined, so that (for example) 'adaptations are traits that are selected for' is a contingent truth rather than a definition. Cf. Skinner once again. We said that the theory of operant conditioning doesn't predict the outcome of split-stimulus experiments. Possible reply: 'Sure it does; for example, it predicts that an animal will always generalize to whatever was the "effective" stimulus property in the learning trials.' That's a bona fide empirical claim if, *but only if*, the explication of 'effective stimulus property' doesn't refer to the outcome of generalization experiments. If a theory specifies the effective stimulus by saying it's the one that controls generalization in split-stimulus experiments, it mustn't predict the outcome of the split-stimulus experiments by saying that generalization will be to whatever property was effective in training. The analogy to natural selection is exact, since, in practice, 'ecological problems' and the like are interdefined with 'adaptations' and the like. An ecological property *just is* whatever some phenotypic trait is an adaptation to; and adaptations *just are* phenotypic traits that solve ecological problems. So 'adaptations are always solutions to ecological problems' isn't after all a bona fide empirical claim; it's just a truism, like 'bachelors always turn out to be unmarried.'

If it hadn't been for the intensionality of 'selection-for', one might have treated outbreaks of free-rider problems as an issue proprietary to the foundations of psychology; one that indicates how tenuous our grasp of the nature of intensional states and processes actually is. Philosophers who never much believed in beliefs or desires (Davidson, Quine, Dennett, etc.) would then have a right not to care much if free-rider problems crop up in cognitive psychology. But now, it seems, we're up to our ears in intensional indeterminacy, not just in folk psychology and cognitive science, but also in the theory of evolution; which has seemed, to many (including Davidson, Quine, Dennett, etc.) to be the very jewel in macrobiology's crown. If there's a 'universal acid',[26] it's not the theory of natural selection, it's the problem of intensionality.

So, as Henry James liked to say, 'here we are'.

Then what kind of theory is the theory of natural selection?

Exasperation speaks: 'Do you guys really want to say that adaptationist explanations aren't ever any good; that selection histories never explain phenotypic traits, psychological or otherwise? Surely you're aware that the textbooks simply teem with good examples to the contrary. These textbook explanations purport to, and often clearly do, give reasons why phenotypes are the way they are; why there are lots of populations of t1s, but few or no populations of t2s. Well, what are we to make of the textbook paradigms of adaptationist explanation if, as you say, adaptationism isn't true but empty?'

We think there are indeed some bona fide adaptationist explanations and that what they are is precisely what they seem to be on the face of them: they're *historical* explanations. Very roughly, historical explanations offer (not laws but) plausible narratives; narratives that purport to articulate the causal chain of events leading to the event that is to be explained. Nomological explanations are about (metaphysically necessary) relations among properties; historical narratives are about (causal) relations among events. That's why the former support counterfactuals, but the latter do not.[27, 28]

Historical narratives are, as far as we know, perfectly OK; certainly they are often thoroughly persuasive. But they don't subsume events under laws, and they therefore don't support counterfactuals. And, as we've been seeing, it's counterfactuals that we need to solve free-rider problems within the adaptationist schema. If adaptationist theories are historical rather than nomological, *that explains why* free-rider problems cannot be solved within the adaptationist framework.

'She fell because she slipped on a banana peel.' Very likely she did; but there's no law – there's not even a statistical law – that has 'banana peel' in its antecedent and 'slipped and fell' in its consequent.[29] Likewise, Napoleon lost at Waterloo because it had been raining for days, and the ground was too muddy for cavalry to charge. (So, anyhow, we're told; and who are we to say otherwise?) But it doesn't begin to follow that there are laws that connect the amount of mud on the ground with the outcomes of battles.

We suppose metaphysical naturalists (among whose ranks we claim to be) have to say that what happened at Waterloo must have been subsumed by some covering laws or other. No doubt, for example, it instantiated (inter alia) laws of the mechanics of middle-sized objects. But it doesn't follow that there are laws about mud so described, or about battles so described – still less about causal connections between them so described; which is what would be required if 'he lost because of the mud' is to be an instance of a law-subsumption kind of explanation or if it were supposed to support counterfactuals about what would have happened if it hadn't rained.

We suppose, likewise, that when a $t1$ creature competes with a $t2$ creature, some laws or other must govern the causal interactions between them. The question, however, is whether they are laws about competitions; or, indeed, whether they are even laws of macrobiology. We don't imagine Darwin would be pleased if it turned out that, although there is indeed an explanation of the mutability of species, it exploits not the vocabulary of competition, selection and the like, but (as it might be) the vocabulary of quantum mechanics.[30]

It's of a piece with the fact that they don't appeal to covering laws that historical-narrative explanations so often seem to be post hoc. The reason they so often seem to be is that they usually are. Given that we already know who won, we can tell a pretty plausible story (of the too-much-mud-on-the-ground variety) about why it wasn't Napoleon. But, what with there being no covering law to cite, we doubt that Napoleon or Wellington or anybody else could have predicted the outcome prior to the event. The trouble is that there would have been a plausible story to explain what happened whoever had won; prediction and retrodiction are famous for exhibiting this asymmetry. That being so, there are generally lots of reasonable historical accounts of the same event, and there need be nothing to choose between them. Did Wellington really win because of the mud? Or was it because the Prussian mercenaries turned up just in the nick of time? Or was it simply that Napoleon had lost his touch?[31] (And while you're at it, what, exactly, caused the Reformation?)[32]

It's not in dispute that competitions between creatures with different phenotypes often differ in their outcomes; and of course, in each case, there must be some explanation or other of why the winner won and the loser did not. But there's no reason at all to suppose that such explanations typically invoke laws that apply to the creatures in virtue of their phenotypic traits. That being so, there need be nothing to choose between claims about the corresponding counterfactuals. Small mammals won their competition with large dinosaurs. But did they do so because of their smallness? That depends (inter alia) on whether they would have won even if there hadn't been a meteor. We can tell you a plausible story about why they might have: small animals are able to snitch dinosaur eggs to eat when the dinosaurs aren't looking (which is bad for the dinosaurs' fitness.) On the other hand, we can tell you a plausible story about why, absent the meteor, the mammals would not have won: there wouldn't have been selection for tolerance to climate change, which the mammals had but the dinosaurs did not. (Notice that, according to the latter story, it wasn't the smallness or quickness of the mammals that was selected for, but the range of temperatures they were able to tolerate.)[33] So, which of the counterfactuals do our evolutionary narratives about the extinction of dinosaurs support? Neither? Both? And, likewise, what trait did evolution select for when it selected creatures that protect their young? Was it an altruistic interest in their offspring or a selfish interest in their genes? Well ...

There is, however, a model of adaptationist explanation that seems to fit the facts pretty well. If it's otherwise viable, it suggests that such explanations, although they aren't nomic or counterfactual supporting, have perfectly respectable precedents. If adaptationist explanations are species of historical narratives, everything can be saved from the wreckage except the notion of selection-for. That's all right because the mechanism of evolution isn't selection-for phenotypic traits. The mechanisms of evolution are the subject matter not of evolutionary theory but of the vignettes that natural history retails case by case. Evolution is a kind of history, and both are just one damned thing after another.

Rhetorical conclusion

Here's an analogy (in fact, we think, it's quite a close one). For each person who is rich, there must be something or other that explains their being so: heredity, inheritance, cupidity, acuity, mendacity, grinding the faces of the poor, being in the right place at the right time, having friends in high places, sheer brute luck, highway robbery, whatever. Which things conduce to getting rich is, of course, highly context dependent: it's because of differences in context that none of us now has a chance of getting rich in (for example) the way that Genghis Khan did; or in the (not dissimilar) way that Andrew Carnegie did; or in the (quite different) way that Andrew Carnegie's heirs did; or in the (again quite different) way that Liberace did; and so forth. Likewise, the extreme context sensitivity of the ways of getting rich make it most unlikely that there could be a theory of getting rich per se, all those how-to-get-rich books that they sell in airports notwithstanding. In particular, it's most unlikely that there are generalizations that are lawful (hence counterfactual supporting, not ad hoc and not vacuous, and so forth)[34] that specify the various situations in which it is possible to get rich and the properties by virtue of which, if one had them, one would get rich in those situations.[35] This is, please notice, fully compatible with there being entirely convincing stories – stories that one *ought* to be convinced by – that explain, case by case, what it was about a guy by virtue of which he got as rich as he did in the circumstances that prevailed when and where he did.

We think adaptationist explanations of the evolution of heritable traits are really a lot like that. When they work it's because they provide plausible historical narratives, not because they cite covering laws. In particular, *pace* Darwinists, adaptationism *does not* articulate the mechanisms of the selection of heritable phenotypic traits; it couldn't because there aren't any mechanisms of the selection of heritable phenotypic traits (as such). All there are are the many, many different ways in which various creatures manage to flourish in the many, many environmental situations in which they manage to do so. Diamond (in Mayr, 2001, p. x) remarks that Darwin didn't just present 'a well-thought-out theory of evolution. Most importantly, he also proposed a theory of causation, the theory of natural selection.' Well, if we're right, that's exactly what Darwin *did not* do: a 'theory of causation' is exactly what the theory of natural selection is not. Come to think of it, it's exactly what we still don't have.

From the viewpoint of the philosopher of science, perhaps the bottom line of all this is the importance of keeping clear the difference between historical explanations and nomological explanations. Just as there is nothing obviously wrong with the former, there is likewise nothing obviously wrong with the latter.[36] Typically, they start with a world in which the initial conditions and the natural laws are specified, and they deduce predictions about what situations will transpire in that world. It's true by definition that the explanation of an event by reference to a law must cite some property of the event in virtue of which the law subsumes it. Nothing has a nomological explanation unless it belongs to a natural kind. (We take what we've just said to be a string of truisms.) Nomological explanations have had a good press in philosophy, and rightly so. Whether or not they are the very paradigms of scientific explanation, it's pretty clear as a matter of fact that many scientific explanations are, or incorporate appeals to, empirical laws.

But nor is there anything wrong with explanations that consist of historical narratives. Roughly, a historical narrative starts with an event for which it seeks to provide an empirically sufficient cause (it was for want of a shoe that the horse was lost). So historical narratives are inherently post hoc (though not, of course, inherently ad hoc). The causally sufficient conditions that historical narratives invoke belong, in familiar ways, to chains of such conditions, which (assuming determinism) can go back as far as you choose (it was for want of a nail that the shoe was lost, and so on). How far back such an explanation ought to go depends, as one laughingly says, on pragmatic factors: what is being explained and to whom, and to what end.

Many paradigm scientific theories are, we think, best understood as historical narratives; consider, inter alia, theories about lunar geography, theories about why the dinosaurs became extinct, theories about the origin of the Grand Canyon, or of the Solar System or, come to think of it, of the Universe. All these projects (and surely many others) are post-hoc searches for chains of sufficient causal conditions whose satisfaction would explain the occurrence of the event in question. If we're right, adaptationist theories about how heritable traits evolve are also of this kind.

That's really just to say that a collection of the various mechanisms of adaptation wouldn't constitute a natural kind; not, at least, if the model for explanation invokes subsumption under nomologically necessary generalizations. But if there are no nomologically necessary generalizations about the mechanisms of adaptation as such, then the theory of natural selection reduces to a banal truth: 'If a kind of creature flourishes in a kind of situation, then there must be something about such creatures (or about such situations, or about both) in virtue of which it does so.' Well, of course there must; even a creationist could agree with that.

What makes the adaptationist literature such fun to read is not the laws of evolution it proposes (it doesn't propose any), and likewise not the mechanisms of phenotypic change that it uncovers (it doesn't uncover any). It's the stories it has to tell about how many strange kinds of creatures there are; and how, case by case, the creatures got to be so strange; and how, strange as they are, each has somehow found a way to make a living. But, to repeat the main moral: from the fact that there are adaptationist explanations that compel rational belief, *it does not follow that there are laws of adaptation*. And if there aren't any laws of adaptation, there is (as far as anybody knows) no way to construct a notion of *selection-for* that isn't just empty. And 'selection-for' is not a notion that a (neo)Darwinian account of evolution can do without.

8

DID THE DODO LOSE ITS ECOLOGICAL NICHE? OR WAS IT THE OTHER WAY AROUND?

The previous three chapters were devoted to arguing that there are serious, perhaps intractable, problems with the theory that natural selection is the primary mechanism of evolution. Because selection-for is intensional (or, if you prefer, because what are selected-for are not creatures but their traits, and the individuation of traits is intensional) there can be coextensive but distinct phenotypic properties, one (but not the other) of which is conducive to fitness, but which natural selection cannot distinguish. In such cases, natural selection cannot, as it were, tell the arches from the spandrels. That being so, adaptationist theories of evolution are unable, as a matter of principle, to do what they purport to do: explain the distribution of phenotypic traits in a population as a function of its history of selection for fitness.

Moreover, this line of argument is contagious; not just selection-for but a whole galaxy of other concepts that adaptationist explanations routinely employ suffer from the same disease. These include, notably, such notions as 'ecological niche', 'problem of adaptation' and 'biological function', all of which are interdefined with 'selection-for' and thus inherit the problems that intensionality occasions.

Consider the adaptationist notion of a 'problem of adaptation'. In familiar and untendentious cases, solutions are identified by reference to the problems that they solve, not the other way around. That's why, although there are many unsolved problems, there are no unproblemed solves; it's why you can lock things up with keys but you can't key things up with locks; and so forth. The order of metaphysical dependence is that keys solve the problem of finding something to open locks, not that locks solve the problem of finding something for keys to open. In adaptationist theory, by contrast, there's a sort of topsy-turvy: whether a feature of the environment constitutes an evolutionary problem for a creature depends on whether the creature's phenotype was selected for solving it. But that there are spiders, who would have guessed that how to spin webs to catch flies is an ecological problem? Or that there are creatures whose fitness is a consequence of their having solved it?

A competition among fish isn't likely to turn on the height of trees on the shores of the pond they inhabit; but a competition among birds may very well do so. It follows that the height of the trees presents a problem of adaptation for the birds but not for the fish. Conversely, if the birds weren't adapted to the height of the trees, that wouldn't show that they had failed to solve one of their problems of adaptation; it would only show that coping with tree heights isn't among the problems of adaptation that their phenotypes evolved to solve. This is a rigged game. The rule is: if a kind of creature fails to solve an evolutionary problem, it follows that that isn't an evolutionary problem for that kind of creature. Quite generally, if a creature fails to fit an ecological niche exactly, it follows that that isn't exactly the creature's ecological niche. The long and short is: if evolutionary problems are individuated post hoc, it's hardly surprising that phenotypes are so good at solving them.

We're stressing this point because there is a quite general kind of argument in favour of the thesis that evolution must consist mostly of adaptation, hence that exogenous variables must be the major factors in phenotypic change. This argument has considerable face plausibility; that it does is perhaps the main reason why so many people think that, whatever challenges to adaptationism may turn up, some or other of its variants must be true. We think that, its face plausibility notwithstanding, this argument is deeply confused.

The exquisite fit

A recurrent motif in adaptationist texts is their emphasis on the close fit of the evolved phenotypic properties of organisms to the ecologies that the organisms occupy. For example (Sober, 1993, p. 186): 'The exquisite fit of organisms to their environments is one of the central phenomena that the theory of evolution by natural selection attempts to explain.' And here's Ernst Mayr:

> How can we explain why organisms are so remarkably well adapted to the environments in which they live? ... Of course a bird has wings to fly with and other attributes that are needed for its aerial existence. Of course a fish has a streamlined shape and fins to enable it to swim ... So it is with all the properties of adapted organisms ... When you begin to think about this deeply, you begin to wonder how this admirable world of life could have reached such perfection. By perfection I mean the seeming adaptedness of each structure, activity and behavior of every organism to its inanimate and living environment.
>
> Mayr, 1963, p. 147

If you have any familiarity with the canonical literature on evolution, you will have encountered a plethora of such passages. Viewed from a Darwinist's perspective, they amount to enthymemic arguments for adaptationism. (An enthymeme is an argument with a 'missing' or 'suppressed' premise.)

Creatures evolve to fit their ecologies; that they do cannot be an accident. It cannot be just good luck when a kind of creature finds itself in a kind of environment in which its kind of phenotype is fit to survive and flourish. Divine solicitude might explain it; everybody knows that God tempers the wind to the shorn lamb. But we are committed to a naturalistic biology, so God is out. What, then, are the naturalistic options?

At first blush, adaptationism seems to make the puzzle go away: creatures fit their ecologies because it's their ecologies that design their phenotypes. As Sterelny and Griffith put it: 'One of the virtues of the received [adaptationist] view is the elegance and simplicity of its picture [of the relation between evolutionary biology and ecology]. Selection shapes organisms to their environment' (Sterelny and Griffith (1999) pp. 48–49).

If you assume that phenotypes vary at random from generation to generation; and if you assume that exogenous, ecological factors are what primarily determine whether a creatures lives or dies; and if you assume that dead creatures have, on average, fewer offspring than living creatures do, then simple statistics guarantees that (all else being equal) evolution will tend towards phenotypes that are fit for the ecologies that they occupy. Notice, however, that this guarantee is on offer only on the assumption that the direction of evolution is sensitive primarily to exogenous factors. If you drop that assumption, then the excellent fit between creatures and their environments is a plain miracle. So, the argument concludes, we had better not drop that assumption. Come what may, we had better cling to an adaptationist account of what causes phenotypes to change. (Modified, to be sure, by lots of other phenomena that everybody acknowledges to be salient: genetic drift, neutral mutations, founder effects, migration, and so forth.)

That's the argument in a nutshell, and many find it thoroughly convincing, indeed decisive. But, for all that, it's fallacious. You don't after all need an adaptationist account of evolution in order to explain the fact that phenotypes are so often appropriate to ecologies, since, first impressions to the contrary notwithstanding, there is no such fact. Or, more precisely, there is no such contingent fact. It is just a tautology that (if it isn't dead) a creature's phenotype is appropriate for its survival in the ecology that it inhabits. Let's see why this argument, convincing as it may sound, is deeply confused.

Environment versus ecological niche

Here's the point: a creature's ecology must not be confused with its environment. The environment that creatures live in is common to each and every one of them – it's just 'the world' (for a rather nuanced notion of environment in evolutionary biology, see Brandon, 1994). By contrast, a creature's ecology consists of whatever-it-is-about-the-world that makes its phenotype viable. That's to say: it is constituted by those features of the world in virtue of which that kind of creature is able to make a living in the world. In effect, the notions 'ecology' and 'phenotype' (unlike the notions 'environment' and 'phenotype') are interdefined. Since they are, it's hardly surprising that a creature's phenotype reliably turns out to be in good accord with its ecology. Do not, therefore, be amazed that the seagull's wings meet with such remarkable perfection the demands that its airy ecology imposes. If seagulls didn't have wings, their ecology wouldn't be airy.

Here's a passage from Mayr (1963) that strikes us as remarkably confused even by the prevailing standards:

> There are two ways to define an [ecological] niche. The classic way is to consider nature to consist of millions of potential niches occupied by the various species adapted to them. In this interpretation the niche is a property of the environment. Some ecologists, however, consider the niche to be a property of the species that occupies it. For them the niche is the outward projection of the needs of a species ... [Many] pieces of evidence show that the classical definition of the niche, as a property of the environment, is preferable. Biogeographers know that every colonizing species has to become adapted to the prospective niches it encounters.

So niches, as Mayr understands them, are of an ontological kind with those headaches sometimes imagined by Oxford philosophers to lurk in a corner of the room, waiting for someone to have them. The picture seems to be: here's the niche, just longing to be filled; with luck, some or other phenotype comes along and fills it. 'It would be quite misleading to say that there are no woodpecker niches in New

Guinea. Actually, the open niches are virtually calling for them' (Mayr, op. cit., p. 152). Were these open niches always there, one wonders? And were the various dinosaur niches in place ('open') even when there were only protozoa? For that matter, are they still open now? Mayr notices, in passing, that his way of talking about niches makes the notion of an environment equivocal. 'The word environment itself is often used in two very different senses, for all the surroundings of a species or biota or only for the niche-specific components' (pp. 152–53); so, in effect, Mayr's Platonic way of individuating niches only moves the lump from under the rug in the hall to under the rug in the parlour: the price of an absolute notion of niche is a relativized notion of environment, with no gain over all. The moral is that, either way, something has to be relativized to the creatures whose phenotypes are to be explained; and, whatever that 'something' is, be it a creature's environment or its niche, it won't be contingent that the creature's phenotype fits it. None of this seems to worry Mayr; but it should.

The main argument of this book so far has been that, because of the intensionality of 'select-for' and 'trait' you can't infer from 'Xs have trait t and Xs were selected' to 'Xs were selected for having trait t'. We now see that the same applies to the individuation of ecological niches and to problems of adaptation. So, for example, you might have supposed that if large-tailed peacocks generally win competitions with small-tailed peacocks, then there is an ecological niche for large-tailed peacocks; and that the reason that peacocks have evolved large tails is that they are thus enabled to fill that niche. But, on second thought, no. Suppose that, in consequence of their selection history, large-tailed peacocks predominate in a population. It doesn't follow that there is, or ever was, an ecological niche 'for' large-tailed peacocks per se. Ecological niches are intensional objects; even assuming that Xs were selected and that Xs are Gs it doesn't follow that there is a niche for Xs that are G. The long and short is: on the one hand, it's interesting but false that creatures are well adapted to their environments; on the other hand, it's true but not interesting that creatures are well adapted to their ecologies. What, then, is the interesting truth about the fitness of phenotypes that we require adaptationism in order to explain? We've tried and tried, but we haven't been able to think of one.

We think that this is important, so we'll say it again: you don't need the theory of evolution to explain why a creature's phenotype is well adapted to its environment (i.e. to the world); that follows simply from the fact that there are creatures with that phenotype. All creatures that are neither extinct nor imaginary are *ipso facto* adapted to the world. But neither is the theory of evolution needed to explain the adaptedness of a creature's phenotype to its ecological niche. Since niches are individuated post hoc, by reference to the phenotypes that live in them, if the creatures weren't there, the niches wouldn't be there either. We take this to be a dilemma for adaptationists. Unless they can deal with it, there would seem to be no 'exquisite fitness' of phenotypes to lifestyles for their adaptationism to explain.

It should be said in praise of Richard Dawkins that he is from time to time sensitive to this sort of worry. In his book *The Blind Watchmaker*, he imagines 'a philosopher' who argues: 'swallows fly but they don't swim; and whales swim but they don't fly. It is with hindsight that we decide whether to judge the[ir] success as a swimmer or a flyer' (Dawkins, 1986, pp. 8–9). Precisely so. But then, how can we avoid circularity if we use the history of its selection-for-fitness to explain the adaptation of a phenotype to its environment? Dawkins has an answer:

> If no matter how randomly you threw matter around, the resulting conglomeration could be said, with hindsight, to be good for something, then it would be true to say that I cheated over the swallow and the whale. But biologists can be much more specific than that about what would constitute being 'good for something'. The minimum requirement for us to recognize an object as an animal or plant is that it should succeed in making a living of some sort … It is true that there are quite a number of ways of making a living … But however many ways there may be of being alive, it is certain that there are vastly more ways of being dead …
>
> Dawkins, 1986, p. 9

Now, we like a good wisecrack too, but this passage really does strike us as perplexing. Consider: 'the minimal requirement for us to recognize an object as an animal or plant is that it should succeed in making a living of some sort.' Well, it's certainly true that the minimal requirement for us to recognize an object as a *living* animal or plant is that it have some way of making a living. But that is surely truistic, since an object that can't make a living is *ipso facto* not alive, and a dead animal or plant isn't an animal or plant *sans phrase*. So the substance of Dawkins's rejoinder must be the claim that 'biologists can be much more specific than that about what would constitute being "good for something"'. But, in fact, can they? There is, to our knowledge, no more an un-question-begging account of 'being good for something' than there is an un-question-begging account of 'being an adaptation'. Each is explicated by reference to the other, so neither is able to stand alone: there is something that a phenotypic trait is 'good for' if, and only if, there is an ecological problem that the trait is selected for solving; there is something that a trait was selected for if, and only if, there is something that the trait is good for (i.e. if, and only if, a creature's having the trait is a way for the creature to make a living). Tweedledee holds up Tweedledum, and vice versa. Sooner or later, all fall down.

As we read him, Dawkins is proposing to individuate niches, evolutionary problems and the like by reference to some independently characterized notion of a creature's way of 'making a living.' But, on second thought, that won't do because there is no such notion. To the contrary, 'makes a living by …' is itself intensional and belongs to the same circle of interdefinition as 'selection-for', 'niche', 'problem of adaptation' and the rest. Since there is no notion of a trait 'being good for something' that doesn't presuppose the notion of an adaptation, the notion of a way of making a living can't be taken for granted in the analysis of other adaptationist notions. If, as Dawkins suggests, biologists have a notion of 'a way of making a living' that isn't relativized to notions like adaptation, ecology and selection-for, they have been strikingly reticent about saying what it is.

Ask yourself how many ways of making a living there are. The question may strike you at first as imponderable, but it's not. To the contrary, the answer is trivially obvious: there must be at least as many ways of making a living as there are actual species and there must be at most as many ways of making a living as there are possible species. That must be so, since a necessary and sufficient condition for a species to be possible is that the creatures that belong to it have some way of making their livings in the ecological niche that they occupy. Accordingly, if a species becomes extinct, then something that used to be a way of making a living has *ipso facto* ceased to be one. The extinction of the dodo was the very same event as the extinction of the dodo's way of making a living, so neither can serve to explain the other.

So then, what becomes of all the stuff about the fine-tuning of phenotypes to the demands that their ecologies impose? In Dawkins's way of putting it, it turns out that there is only one demand that an ecology makes on is inhabitants: 'that [they] should succeed in making a living *of some sort* (more precisely that [they], or at least some members of [their] kind, should live long enough to reproduce)' (Dawkins, op. cit., p. 9). To this question, every creature that is alive must have found an answer; *that is true by definition* since, at a minimum, its progenitors must have lived long enough to produce it.

Of course creatures fit their niches with great precision: if a certain species fails to occupy a certain niche exactly, all that follows is that that isn't exactly the niche that the species occupies. Imagine a research programme directed to explaining why each creature fits so perfectly its proprietary hole in space. Would the National Science Foundation be well advised to fund it? (We wouldn't put it past them.) Or imagine Scrooge before his tragic capitulation: 'The man who is living in the gutter on scraps from the tables of the rich has nothing to complain of; for, so long as his reproductive capacity is unimpaired, it follows that he is perfectly adapted to making a living in exactly the way that he does; viz. in the gutter, on scraps from the tables of the rich.' That would be a joke if it were funny.

To sum up so far: although it's very often cited in defence of Darwinism, the 'exquisite fit' of phenotypes to their niches is either true but tautological or irrelevant to questions about how phenotypes evolve. In either case, it provides no evidence for adaptationism. That being so, you might wonder what kind of theory does explain why there are the phenotypes there are. This is a question we'll return to in Chapter 9, but we want to say a preliminary word or two here.

Natural selection theory is often said to provide a mechanism for the evolution of phenotypes. That, however, is precisely what it doesn't do. What explains why there are the phenotypes there are is not natural selection but natural history. It is, as we've been seeing, just a truism that birds are adapted to their airy ecology. But what isn't a truism is that the bird's wings are the mechanism of this adaptation. If, in the ecology they occupy, birds with wings are better off than birds without them, there must be something about the birds, or about the ecology, or about the two together, in virtue of which birds with wings are better off in that ecology than birds without them. That's just a routine application of the principle of sufficient reason; as such it's true a priori and applies, sight unseen, to birds that have wings, fish that have gills, germs that are resistant to penicillin, and so forth indefinitely. So, as one politician asked about another politician, 'where's the beef?'

The beef comes not from adaptationism but from the details of natural history. You have to look into the structure of niches (how birds manage to fly; how fish manage to breathe under water; whatever). That's how to find out how having wings conduces to the fitness of birds. It's natural history that gets you out of the circle that plagues selection theory, in which 'niche' is defined in terms of 'adaptation' and 'adaptation' is defined in terms of 'niche'.

The intensionality of contexts such as 'makes a living by ...' undermines the claim that the theory of natural selection supplies the mechanism of evolution; but it is fully compatible with the explication of such mechanisms by natural history. What happens in practice is that scientists have not just their theories to go by but also their background sophistication and their noses; and it's often (though by no means always) obvious at a glance how some of a creature's traits affect its viability in a certain environment while other of its traits do not. That sharks have teeth clearly matters to how they manage their affairs – even though exactly how many teeth they have very probably does not. This is all clear on the face of it to anybody who has had anything to do with sharks. It's having to do with sharks that tells us what the shark's teeth are good for; the theory of evolution says nothing whatever on the topic. What the scientist can tell, at a glance or otherwise, is one thing, what the scientist's theory can tell is quite another.

Two things should be noted about natural histories (in contrast to theories of evolution such as adaptationism proposes). In the first place, natural histories are invariably post hoc. Lacking observations of spiders, nothing (least of all the theory of natural selection) could have predicted that there are creatures that have the spider's kind of adaptation to their niches. What happened is that somebody who knew that spiders make a living by eating flies looked carefully at their natural history and was thus able to figure out that spinning webs is how they do so. Unlike selection theory, natural history is as far as you can get from being generic; it's packed with details, many of which are distinctly surprising and more than a few of which are distinctly gruesome.

Second (or maybe it's just the first point seen from a different perspective), natural histories tend to differ wildly from creature to creature. There's a story about how spiders catch flies to eat, and there's a story about how oak trees distribute their seeds, but the two have little or nothing in common; there aren't, as a philosopher might put it, laws – or even reliable empirical generalizations – about their mechanisms of adaptation or the structure of their niches. Some of them work in one way, others work in quite different ways, and no two are likely to work in much the same way. Natural history is a lot like history *tout court*; it's about what actually happened. There are, often enough, interesting things to say about what actually happened; but none of them is counterfactual supporting; none them is nomologically necessary; and all of them are post hoc. That is, none of what there is to say about what actually happened amounts to a theory of history.

'Well, then, isn't there anything that follows from natural selection theory that is both general, and contingent and not post hoc?' Actually, we think there is: namely the claim that the forces that form phenotypes over time are, by and large, exogenous. But although that's general, and contingent and not post hoc, it isn't at all clear that it's true. There is recently a noticeable flowering among biologists of the idea that the contribution of endogenous forces may be decisive in explaining how phenotypes evolve (see Chapter 2).

Darwin was interested in the question of why there are species; that is, why phenotypes fall into families when they are grouped by similarity. He thought that the similarities among species is largely the effect of their common ancestries and, occasional counter-examples notwithstanding, he was certainly right about that. But there's a different question, one that Darwin didn't ask, the pursuit of which might have proven – indeed might still prove – equally fruitful; namely, why certain perfectly imaginable phenotypes simply don't occur, either here and now or anywhere in the fossil record. Why, for example, aren't there pigs with wings? Surely the answer doesn't lie in an appeal to exogenous selection for fitness. Nobody thinks that if there aren't pigs with wings it's because the winged pigs were selected against in their prehistoric competitions with wingless ones. Rather, pigs don't have wings because there is no place on pigs to put them. There are all sorts of ways you'd have to change a pig if you wanted to add wings. You'd have to do something to its weight, and its shape, and its musculature, and its nervous system, and its bones; to say nothing of retrofitting feathers. Once a kind of creature is well on the evolutionary path to being a pig there's no way for it to add wings to its phenotype. None of this, however, has much to do with the operation of exogenous forces. Often, what explains why some phenotype doesn't occur is not that selection operated against it, but that some arrangements of the phenotypic bits and pieces are not possible. There are constraints on phenotypes that, as it were, operate not from outside but from below.

How many such endogenous constraints are there, and of what kinds? Nobody knows exactly, nor do we claim to. It is clear, however, that the more endogenous constraints there are, the less work the ecology has to do in shaping phenotypes. At the limit, there isn't any work at all for it to do; at the limit, creatures are the way they are because they couldn't be any other way. If, at a certain stage of evolution, the only endogenously possible phenotypes are large dinosaurs and small mammals, then selection doesn't have much to do to determine the course of evolution from that stage; all that's required is to hit the dinosaurs over the head with a meteor.

We'll return to this and related lines of thought in the last chapter. Suffice it for now that:

(1) It doesn't need explaining that organisms are 'exquisitely adapted to their ecologies'; insofar as it's true that they are, it is truistic that they are.

(2) It does need explaining what it is about a creature's phenotype in virtue of which it is able to make a living in the ecology it inhabits. Such explanations are in the domain of what one might call 'synchronic natural history'; evolutionary theories don't even purport to provide them. If you want to understand wherein the adaptivity of a creature's phenotype consists, ask the (synchronic) question 'how does it make a living?', not the (diachronic) question 'how did it come to make a living in that way?'

(3) It does need explaining how – that is, by what actual historical process – creatures came to have the phenotypes they do. This is the domain of what one might call 'diachronic natural history', of which adaptationism is one version. But there is no obvious reason to suppose, as proponents of natural selection do, that phenotypic properties are by and large the effect of adaptation to exogenous variables.

You win a few, you lose a few: sometimes the fault is in your stars; sometimes it's in yourself, and sometimes you get hit on the head by a meteor. Stuff happens, as one well-known dinosaur remarked. Hence Gould's suggestion that if you ran the tape of evolutionary history a second time, you'd quite likely come out somewhere different. History (natural history included) is about what actually happened; it's not about what *had to* happen; or even about what would happen if Mother Nature were to try again. What had to happen is the domain of theory, not of history; and there isn't any theory of evolution.

9

SUMMARY AND POSTLUDE

'OK; so if Darwin got it wrong, what do you guys think is the mechanism of evolution?' Short answer: we don't know what the mechanism of evolution is. As far as we can make out, nobody knows exactly how phenotypes evolve. We think that, quite possibly, they evolve in lots of different ways; perhaps there are as many distinct kinds of causal routes to the fixation of phenotypes as there are different kinds of natural histories of the creatures whose phenotypes they are (see previous chapter).

But we do have a scattering of remarks to make in aid of gathering some threads together and of suggesting some ways of thinking about theories of evolution that differ from the current adaptationist consensus.

Chapter 1 discussed some striking parallels between the natural selection account of the evolution of phenotypes and the learning-theoretic account of the acquisition of operant behavioural repertoires. Both propose a view of organisms as random generators of traits (phenotypic traits in one case, operant behaviours in the other); both think of the structures they purport to explain as shaped primarily by processes of exogenous selection; and both confront prima facie objections about free-riders. In natural selection these take the form of arch/spandrel problems: How can natural selection distinguish between, on the one hand, phenotypic traits that affect fitness and, on the other hand, their endogenously linked phenotypic correlates? In conditioning theory they take the form of puzzles about 'what is learned': How can reinforcement distinguish the 'effective' stimulus (or response) from correlated stimulus (or response) properties that are merely redundant.[1] We further suggested that, in both natural selection and operant conditioning theory, the free-rider problems are instances of a more general malaise: neither selection nor conditioning can apply differentially to coextensive properties. If properties P_1 and P_2 are coextensive, selection of the first is willy-nilly selection of the second; and conditioning of one is willy-nilly conditioning of both. Finally, again in both cases, failures to cope with free-rider problems are plausibly viewed as failures to support relevant counterfactuals. To say that natural selection can't distinguish between coextensive phenotypic traits is to say that it can't predict what *would be* the relative fitness of a phenotype that had one trait but lacked the other; it can't tell arches from spandrels. To say that operant conditioning theory can't distinguish the effective stimulus from its redundant correlates is to say that it can't predict the outcome of 'split stimulus' experiments.

Past this point, however, the symmetry fails. When Skinner's learning theory came unglued, the appropriate reaction was to abandon his behaviourism. Thus, according to 'cognitive science', learning is mediated by mental representations; and representations *can* distinguish among coextensive properties. If P_1 and P_2 are coextensive, then to reinforce a P_1 stimulus is to reinforce a P_2 stimulus, and vice versa. But even when stimuli are coextensive, a mental representation of a stimulus *as* P_1 is not thereby a mental representation of the stimulus *as* P_2. 'Mentally represent as …', unlike 'reinforce …', is intensional. Accordingly, a psychology that recognizes mental representations can, at least in principle, distinguish the effective stimulus (or response) from its redundant background. Once this had been noticed, the cognitive science research agenda became all but inevitable: explain a creature's behaviour by a theory that specifies how the creature mentally represents the relevant stimulus and response contingencies.

However, the corresponding treatment is out of the question in the case of evolutionary theory. To be sure, introducing mental states into the operation of natural selection would allow it to reconstruct the distinction between *selection* and *selection-for*; and, as we've been seeing, that is just the distinction that a satisfactory treatment of free-rider problems requires. But the cost would be catastrophic. Mental processes require minds in which to happen. So to allow them in the theory of evolution would mean committing precisely the error of which we've been accusing Darwin: construing natural selection on the model of selective breeding. It ought to be common ground among naturalists that evolution is *not* an intentional process; it isn't run by Mother Nature, or by Selfish Genes or by the Tooth Fairy, or by God. Selective breeding is something that somebody *does*. But natural selection is not; it is something that just happens.

So we're confronted with a problem that we've thus far skirted. We need *selection-for* to distinguish traits that affect fitness from mere free-riders. But we can't (indeed, mustn't) suppose that natural selection-for is the result of mental process; the path that proved so fruitful in psychology is closed to us here. The sum and substance is that, although evolutionary biology looks like it's an intentional science; it perfectly clearly can't be. So now what? This begins to sound like a serious dilemma; but in fact there's a way out – to which we now turn.

Mental states are typically *intentional*, hence typically *intensional*. Even if all and only the woolly sheep are stupid, we can distinguish breeders who select sheep for their wool from breeders who select them for their stupidity. But intentional contexts aren't the only ones that work this way; 'nomological' contexts do too. For example, it might be *true but not lawful* that woolly sheep are stupid; it might be true *just by chance*. Compare 'all sheep are vertebrates' which is, presumably not just true but *nomologically necessary*; that is, it's true because it expresses a natural law.

There are various more-or-less equivalent ways of drawing this distinction; one might say that, even if all woolly sheep are stupid in the *actual* world, still there are (nomologically) *possible* worlds in which woolly sheep are smart. By contrast, since it's a law that sheep are vertebrates, there are no (nomologically) possible worlds in which sheep lack backbones. Or one could say that the difference between laws and mere accidentally true generalizations is that the former support counterfactuals but the latter do not. If it's a law that woolly sheep are stupid, it follows that if there were woolly sheep in the attic, there would be stupid sheep in the attic. This way of putting it is germane to the present interests because, as we've been seeing, problems about free-riders *are* problems about counterfactuals. If selection were for arches (rather than for spandrels), then, all else being equal, there *would be* arches even if there *weren't* spandrels. That's so even on the assumption that, in all actual cases, selection *of* one is selection of the other.

The upshot is that, so far at least, we might suppose that the generalizations of natural selection support counterfactuals, *not* because natural selection is an intentional process, but because some true generalizations about natural selection are laws. Among its other virtues, this sort of account would explain why breeding isn't after all a good model for natural selection: when counterfactuals about breeding are true, what supports them is the intentions and preferences of the breeder. But, according to the present suggestion, when counterfactuals about natural selection are true, what supports them are natural laws.

So maybe we could, after all, have exactly what a naturalistic natural selection requires: *intensional* selection without *intentional* selection. Given that, we could presumably also have the distinction between phenotypic properties that are selected-for and mere free-riders; and we can do so without the postulation of God, Mother Nature, Selfish Genes, the Tooth Fairy or any other agent of natural selection. Natural selection doesn't (of course) *have* an agent; but it can distinguish among coextensive phenotypes all the same, since coextensive properties may differ in respect of the laws that subsume them. So the story might well be supposed to go.

But there are no free lunches, and we've already seen why this sort of proposal is unlikely to work. The problem is that it's unlikely that there *are* laws of selection. Suppose that P_1 and P_2 are coextensive but that, whereas the former is a property that affects fitness, the latter is merely a correlate of a property that does. The suggestion is that all this comes out right if the relation between P1 and fitness is lawful, and the relation between P2 and fitness is not. But, as we argued at length in Chapter 7, it's just not plausible that there are laws that relate phenotypic traits per se to fitness. What (if any) effect a trait has on fitness depends on what kind of phenotype it is embedded in, and what ecology the creature that has the trait inhabits. This is to say that, if you wish to explain the effects that a phenotypic trait has on a creature's fitness, what you need is not *its history of selection* but its *natural history*. And natural history offers not laws of selection but narrative accounts of causal chains that lead to the fixation of phenotypic traits. *Although laws support counterfactuals, natural histories do not;* and, as we've repeatedly remarked, it's counterfactual support on which distinguishing the arches from the spandrels depends.

Comparison with psychology is once again illuminating. It really does seem plausible that there are laws about intentional states and processes. What these laws are, and how deep they run, and what counterfactuals they support, are all open questions that only empirical research can answer; as is, for that matter, the whole claim that intentional explanations are the appropriate model for psychological theories. Still, although it doesn't seem likely that there are laws of selection, it does seem that there are laws of intentional psychology.

For a scattering of examples: it is very plausibly counterfactual supporting that (all else being equal) Necker cubes are seen as ambiguous; and that working memory is largely item limited; and that apparent brightness varies as a logarithmic function of the intensity of the illumination; and that arguments that turn on contraposition are harder to assess than arguments that turn on *modus ponens*; and that stereotypic instances of a kind are easier to recognize than marginal instances; and that concepts that subtend 'middle-sized' objects are learned earlier than concepts that express abstract objects; and that free recall of nonsense stimuli generates a serial position curve …; and so forth through many, many textbook examples – whereas, as we remarked in Chapter 7, it's hard to think of even one plausible example of a law of the form 'P1 phenotypes are more likely to be fit than P2 phenotypes.' The point isn't, of course, that mental phenomena are subsumed by laws but the fixation of phenotypes is not. Quite the contrary: if determinism is true, *everything that happens* is subsumed by laws; and determinism is (we suppose) common ground in the present discussion. Rather, the difference appears to be that intentional psychology constitutes a *level of explanation*, but evolutionary biology does not.

To a first approximation, mental phenomena are explained by reference to their relations to other mental phenomena. Here again there is a plethora of likely examples: *how one believes things to be* depends on *how things appear to one to be* (and/or on *what other things one believes*); *what one does* depends on *what one decides to do*, and *what one decides to do* depends on *what one wants* and *what one believes*; *what one attends to* depends on *what one's interests are*; *what one sees* depends (often enough) on *what one expects to see*; what one expects to see depends (often enough) on one's prior beliefs; and so forth.

It would seem plausible, in short, that there is an intensional level[2] of psychological explanation. It seems reasonably likely that the ontology of that level consists of mental objects, states and processes; either what happens is so constituted, or it isn't in the domain of (cognitive) psychology. Another way to put it is that cognitive psychology is, or anyhow purports to be, a one-level theory; and phenomena at level are characteristically intentional.

What, then, about theories of evolution? Adaptationism, as we read it, is also a one-level theory: it purports to explain the fixation of phenotypic properties as the effects of selection by ecological variables. Processes that aren't subsumed by generalizations at this level aren't in the domain of adaptationist explanations, although they may, of course, provide the mechanisms by which adaptation is implemented, much as processes at the neurological level may be supposed to provide the mechanisms whereby psychological processes are implemented. Likewise, genetic processes and the variety of developmental regulatory mechanisms we have summarized in Part one aren't specified by theories of selection, but one can bet that they are at the source of what *can* become an adaptation.[3]

What's essential about adaptationism, as viewed from this perspective, is precisely its claim that there is a level of evolutionary explanation. We think this claim is just plain wrong. We think that successful explanations of the fixation of phenotypic traits by ecological variables typically belong not to evolutionary theory but to natural history, and that there is just no end to the sorts of things about a natural history that can contribute to explaining the fixation of some or other feature of a creature's phenotype. Natural history isn't a theory of evolution; it's a bundle of evolutionary scenarios. That's why the explanations it offers are so often post hoc and unsystematic.

It's in the nature of explanations in natural history to collapse across ontological levels. Maybe what determines some aspect of a creature's phenotype is the local weather; or the composition of the local atmosphere; or the salinity of the local water; or something about the creature's biochemistry; or something about the biochemistry of the creature's prey; or maybe the creature's relative size; or its relative weight; or its buoyancy; or the colours of the things in its environment; or the density with which its environment is populated; or the anatomy of its birth canal; or maybe it's some aspect of the cosmic radiation that the creature is subject to …. Or maybe it's all of these acting at once. In short, practically anything about its macrostructure or its microstructure or its internal environment or its external environment can play a role in the fixation of a creature's phenotype. If, putting it in Dennett's terms, natural selection is a theory about what a creature's phenotype 'carries information about' (viz. it's the theory that a creature's phenotype carries information about its ecology) then the alternative we're commending is that it carries information about the creature's natural history,[4] and the features of a creature's natural history about which its phenotype carries information are just about arbitrarily heterogeneous. Natural history is just one damned thing after another. This should seem, on reflection, unsurprising since, to repeat, natural history is a species of history, and history is itself just one damned thing after another.

Marx and many other nineteenth-century luminaries notwith-standing, there is no *level* of historical explanation; a fortiori, there is no theory of history. Rather, history is composed of many, many causal chains, the links of which are wildly various, not just from the perspective of basic sciences like physics, but also from the per-spectives of the special sciences. 'For want of a nail, the shoe was lost; for want of the shoe, the horse was lost; for want of the horse, the rider was lost; for want of the rider, the message was lost; for want of the message, the battle was lost', and so forth. That may well explain the loss of the battle. But it's not couched in the vocabulary of any of the sciences; nor do the transitions from event to event that it cites instance laws (or even reliable empirical generalizations) that any of the sciences articulates, or is likely to. On the present view, Darwin made the same sort of mistake that Marx did: he imagined that history is a theoretical domain; but what there is, in fact, is only a heterogeneity of causes and effects.

By contrast, Skinner was right about there being a level of psycho-logical explanation. What he was wrong about was the explanatory domain of theories at that level: he thought it consisted of instances of stimulus–response relations, whereas it turns out to consist of instances of relations among intentional states and processes. Dar-win's mistake was, however, much deeper than Skinner's. Skinner was wrong about what kind of theory psychology would turn out to be. But Darwin thought that there is a level of explanation where, in fact, there are only bundles of causal histories. Supposing that historical phenomena constitute a theoretical domain is, as we suggested, a very characteristic nineteenth-century mistake; perhaps it's unsurprising that Darwin made it.

That, however, is only half of the story we've been telling you. The patient reader will recall that its inability to solve problems about free-riding (more generally, problems about coextensive phenotypic or ecological properties) was only one of the complaints we've had against natural selection. The other is that natural selection badly underestimates the significance of endogenous factors in the determi-nation of phenotypes: we think the thesis that organisms are random generators of phenotypes can't be sustained even as a first approximation to an explanation of why there are the phenotypes that there are.

Part one reviewed some of the empirical considerations that militate against claiming that organisms are random generators of phenotypes. Part and parcel is that natural selection offers no insight into why there aren't the phenotypes there aren't (why there are, in the terminology of Chapter 5, 'holes' in the biological space of forms). There aren't, and there never were, pigs with wings. That there aren't and weren't needs to be explained; but the explanation surely cannot be selectionist. Mother Nature never had any winged pigs to select against; so pigs not having wings[5] can't be an adaptation. We think such considerations strongly suggest that there are endogenous constraints – quite possibly profound ones – on phenotypes. As far as we can tell, this is slowly becoming the received view in evolutionary biology (see chapters 2–4).

If, however, it is wrong to suppose that phenotypes are generated at random, it is at least equally wrong to suppose that the various correspondences between ecologies and phenotypes are anything like fortuitous. That's what Darwin got right; and it's what he thought that claiming that phenotypic traits are, by and large, adaptations would explain. We do think that the usual rhapsodies about the 'exquisite', 'perfect', 'wonderful', 'amazing', and so forth, fit between creatures and their environments are distinctly overblown; much of what this sort of rhetoric insists on is an artefact of the circular interdefinition of 'ecologies' and 'phenotypes' (see Chapter 8). But we agree it would be perfectly mad to doubt that phenotypes quite often 'carry information' about the environments in which creatures evolved. What we doubt is that the attempt to subsume the aetiologies of phenotypes under a uniform theory is well advised. Just as the pursuit of natural history would seem to suggest, the sources of matches between organisms and their environments are thoroughly heterogeneous. Darwin thought that ecological selection for fitness uncovers the underlying similarity of all – or anyhow most – such cases, but he was wrong. Either adaptationist theories cannot support relevant counterfactuals about trait selection or they draw uncashable cheques on key notions (such as ecology and phenotype), which on close inspection turn out to be interdefined. We suspect that, to a first approximation, the natural history of phenotype fixation really is just about as anecdotal as it seems to be. Familiar claims to the contrary notwithstanding, Darwin didn't manage to get mental causes out of his account of how evolution works. He just hid them in the unexamined analogy between selection by breeding and natural selection. So Darwin didn't after all elucidate the mechanisms by which phenotypes are constructed.

Here's a metaphor that we prefer to Darwin's: organisms 'catch' their phenotypes from their ecologies in something like the way that they catch their colds from their ecologies. The aetiological process in virtue of which phenotypes are responsive to ecologies is more like contagion than selection. There are at least two respects in which this is so: one is that what diseases a creature catches depends not just on what kind of world it inhabits but also, and probably ineliminably, on features of its endogenous structure: features which it may have innately, or may have acquired in consequence of its prior interactions with its ecology. Paramecia don't catch colds, and our catching one cold doesn't prevent us from catching another one. (Compare smallpox.) Both facts need to be explained; but there is no reason at all to suppose that these phenotypic traits were selected-for or that they carry information about the ecologies in which either we or paramecia evolved. (Presumably what they do carry information about is their respective physiologies.)

Second, contagion depends, quite possibly ineliminably, on factors that work at very many different levels of organization; and so, we think, does the fixation of phenotypes. Part of the story about what happens when one comes down with a cold concerns the microstructure of the pathogens and of one's immune system. Part of the story is about what having the virus does to the mucous membranes. And part of it is about the age, sex, health, degree of exposure and so forth of the host. Clusters of facts of all of these sorts (and no doubt of many others) contribute to the explanation of how and why we catch the colds we do when we do. What's surprising isn't that some empirical explanations turn out to be just causal histories; what's surprising is that not all of them do.

Surely, some sorts of interactions between organisms and their environments are causally implicated in the evolutionary fixation of some phenotypic traits; if that weren't so, it really would be miraculous that there are reliable correspondences between the two. But there's no obvious reason to doubt that these interactions are simultaneously structured at many levels of analysis; and it's entirely possible that the story about the aetiology of organism/environment matches may differ from one kind of phenotypic trait to another. If so, then the right answer to 'What is the mechanism of the fixation of phenotypes?' would be: 'Well, actually there are lots.' We repeat that this is not in the least to suggest that the fixation of phenotypes is other than a deterministic, causal and lawful process through and through. No Tooth Fairy need apply. What is denied, however, is that there is a unitary theory (e.g. a unitary theory of organism–environment interactions) in terms of which most or all such phenomena are explained; or that the various kinds of accounts that explain them generally imply that there are laws of exogenous selection.

Perhaps that strikes you as not much; perhaps you would prefer there to be a unified theory – natural selection – of the evolutionary fixation of phenotypes. So be it; but we can claim something Darwinists cannot. There is no ghost in our machine; neither God, nor Mother Nature, nor Selfish Genes, nor the World Spirit, nor free-floating intentions; and there are no phantom breeders either. What breeds the ghosts in Darwinism is its covert appeal to intensional biological explanations, which we hereby propose to do without.

Darwin pointed the direction to a thoroughly naturalistic – indeed a thoroughly atheistic – theory of phenotype formation; but he didn't see how to get the whole way there. He killed off God, if you like, but Mother Nature and other pseudo-agents got away scot-free. We think it's now time to get rid of them too.

10

AFTERWORD AND
REPLY TO THE CRITICS

On our book and its reception

The hardcover edition of this book was first published early in 2010. It was intended to raise two objections to the Theory of Natural Selection (TNS) and to explore their connections to each other and to familiar questions about evolution. First, we claimed that TNS is committed to an untenable externalism: Like Skinner, Darwin held that paradigm explanations of biological (and psychological) structure should invoke relations between organism and their ecologies. But, whereas Skinner's externalism was largely motivated by his methodological commitment to behaviorism, Darwin's was quite different; Darwin held that externalism is the price one pays for adaptationism: only an externalist theory could explain why the features of a creature's phenotype are so often well-adapted to the features of its ecology. The explanation on offer is that phenotypes are shaped by the ecological features to which they are adapted. We suggested, by contrast, that the appearance of adaptation is in large part illusory. The reason a creature's phenotype seems well-adapted to its ecology is that *by definition*, an 'ecological feature' is one to which the fitness of phenotypic traits is sensitive; and a 'phenotypic trait' is *by definition*, one that effects a creature's fitness in relation to its ecology. We aren't, of course, the first to suspect that there are vicious circularities lurking at the heart of TNS. But we have tried to make them explicit, and to document a variety of recent empirical findings that strongly suggest the crucial role of endogenous variables in the evolution of phenotypes. About half of our book is devoted to doing so.

The second problem we raised for TNS has, to our knowledge, hardly been noticed elsewhere in the literature: the tension between its treatment of *selection* and its treatment of *selection-for*. TNS holds, in effect, that though what get *selected* are kinds of creatures (kinds of creatures are what flourish, or fail to, in a given ecology), what creatures get *selected-for* are certain of their phenotypic traits (viz those phenotypic traits that cause their fitness). Problems arise because, unlike selection, selection-for is a paradigmatically *intensional* concept: it is perfectly possible that there should be selection-for one, but not the other, of two coextensive phenotypic traits. The intensionality of selection-for is duly inherited by a variety of other notions that are interdefined with it, and to which TNS is committed. These include, in particular, the notion of a phenotypic trait itself (since one but not the other of coextensive phenotypic traits may be selected-for). This we suggest, is the logical consideration from which the notorious problems about 'arches and spandrels' eventually arise. We argue that because selection-for is intensional and selection is not, TNS can't, even in principle, decide which of its traits is selected for when a kind of creature is selected. This should hardly be surprising; there is an exactly parallel situation in cognitive psychology, where the intensionality of the 'propositional attitudes' – beliefs, desires, and the like – offers a prima facie objection to the naturalisability of 'representational' theories of mind. That there is this previously widely ignored analogy between the (putative) intensionality of mental processes and the (putative) intensionality of evolutionary processes is one of the things that makes the present issues philosophically interesting.

Our claim is that, given coextensive phenotypic traits, TNS can't distinguish ones that are causally active from ones that aren't. Many of the objections that have been raised against us seem unable to discriminate this claim from such quite different ones that we didn't and don't endorse, such as: when traits are coextensive, there is no fact of the matter about which is a cause of fitness; or, when traits are coextensive, there is no way to tell which of them is a cause of fitness; or when traits are coextensive science cannot determine which is a cause of fitness, etc. Such views are, we think, preposterous on the face of them; we wouldn't be caught holding them. To the contrary, it is precisely because there *is* a fact of the matter about which phenotypic traits cause fitness, and because there is no principled reason why such facts should be inaccessible to empirical inquiry, that the failure of TNS to explain that which distinguishes causally active traits from mere correlates of causally of active traits, shows that something is seriously wrong with TNS.

We were, on balance, very pleased the way our book turned out. It seemed to us quite plausible, in the light of the considerations it raised, that TNS is simply untenable and that, insofar as current evolutionary theory presupposes it, current evolutionary theory is due for a thorough reconsideration. We thought of this as a real scientific advance; the next best thing to finding out what one ought to believe is finding out what one ought not. We didn't exactly expect to be awarded a tickertape parade, of course, but we were looking forward to at least a few warm congratulations. In the event, however, the book was received very badly. Almost (though not quite) all the reviews were hostile and some were hysterical. Our arguments and our conclusion were both widely and wildly misrepresented. Many suspected that we are covert Theists, committed to undermining the foundations of the Scientific World View (of which they took themselves to be the anointed custodians). Others reproached us for having opinions on issues that are proprietary to members of the Guild of Professional Biologists. The blogs, in particular, were ablaze with anonymous contumely. Well, what did we expect? Hadn't we heard there's a Culture War on?

Some of the objections we've seen strike us as too silly to bother refuting. Others deserve serious replies. The latter should be addressed at length; they will be in future publications. But there were a number of criticisms that can be replied to succinctly; hence the present update. We propose to quote, and rebut, a scattering of short passages from reviews of our book. Hope springs eternal, so we're told. We hope, at a minimum, to clear the ground for more extended discussions.

When some biologists (indirectly) agree with us

Several reviewers have suggested that we don't know enough about biology to criticize a theory that so many biologists hold dear. The implication is: only someone improperly educated could say the sort of things we do. But we don't think our critics are well-advised to insist on our lack of credentials. For one thing, several of them aren't biologists either. For another, it's a self-defeating line of argument; do they hold that only theologians are licensed to discuss the existence of God?

Everybody makes mistakes, even biologists, even biologists who agree with one another, even great biologists like Darwin. If you think somebody has made a mistake, then it's a good thing for you to say so, so that s/he (or you) can be corrected. Surely that is common ground among scientists, philosophers and everybody else who cares about distinguishing the true from the false. The parochial is the enemy of the true, and should not be encouraged. But we won't go on about this; it's a little embarrassing even to have to mention it. Instead, we report verbatim some recent passages by fully qualified evolutionary biologists, each of whom has earned a PhD from an accredited institution of higher learning, and all of whom are explicit in maintaining that neo-Darwinism (the new synthesis) is gone.

> In the post-genomic era, all major tenets of the modern synthesis have been, if not outright overturned, replaced by a new and incomparably more complex vision of the key aspects of evolution. *So, not to mince words, the modern synthesis is gone.* What comes next? [...] a postmodern state [...]. Above all, such a state is characterized by the pluralism of processes and patterns in evolution that defy straightforward generalization. (our emphasis)
>
> Eugene V. Koonin, Senior Investigator, National Institutes of Health,
>
> 2009 (a)

Evolutionary-genomic studies show that natural selection is only one of the forces that shape genome evolution and *is not quantitatively dominant, whereas non-adaptive processes are much more prominent than previously suspected.* (our emphasis)

Koonin, E. V., 2009 (b)

Although 2009 will be marked by a plethora of celebrations on the subject of evolution, most of the attention is being bestowed on the personalities and historical circumstances surrounding the theory of natural selection, as if this and its synthesis with genetics in the first decades of the twentieth century marks the culmination of the theory of evolution. It does not.' [...] Dogmatic thinking has prevailed all too often in our account, with disastrous consequences for the progress of the fields of microbiology, molecular biology, and the study of the evolutionary process. It led to the stagnant and scientifically invalid notion of the prokaryote; it led to the redefinition of the problem of the gene; and through *a slavish adherence to the modern evolutionary synthesis*, it led to a premature declaration of victory in the struggle to understand the evolutionary process [...] The study of evolution is poised to cast off a century of dogma and to become a true science, fully integrated with discoveries that owe their roots to microbiology and molecular biology. It is time for biology to put its past behind and begin rethinking the discipline's future. *It can no longer afford to keep the study of evolution within the narrow confines of the so-called modern evolutionary synthesis.* (Our emphasis)

Carl R. Woese, Microbiologist, University of Illinois, winner of the 2000 National Medal of Science, and Nigel Goldenfeld, Professor of Physics at the University of Illinois at Urbana-Champaign and Head of the Biocomplexity Group at the University's Institute for Genomic Biology, 2009

Despite elaborate Neo-Darwinist mathematical models that focus on inherited variation in animals, evidence continues to mount that the branches of 'the tree of life' do not just bifurcate. They

do not simply diverge by gradual accumulation of random mutations. Rather, lineages converge, as the result of gene transfers,
mergers, fusions, partnerships, anastomoses and other forms of
alliance. The most accurate modern taxonomies recognize that
Archaebacteria and Eubacteria have become subkingdoms of the
prokaryotes whereas all nucleated organisms (eukaryotes) evolved
symbiogenetically.

> Lynn Margulis (Distinguished University Professor of
> Geosciences at the University of Massachusetts, winner of the 1999
> Presidential Medal of Science) and Michael J. Chapman (Marine
> Biological Laboratory, Woods Hole, MA), 2010

There is a growing appreciation among evolutionary biologists
that the rate and tempo of molecular evolution might often be
altered at or near the time of speciation, i.e. that *speciation is in
some way a special time for genes.* Molecular phylogenies frequently reveal increased rates of genetic evolution associated with
speciation and other lines of investigation suggest that various
types of abrupt genomic disruption can play an important role in
promoting speciation via reproductive isolation. *These phenomena are in conflict with the gradual view of molecular evolution
that is implicit in much of our thinking about speciation and in
the tools of modern biology.* This raises the prospect of studying
the molecular evolutionary consequences of speciation per se and
studying the footprint of speciation as an active force in promoting genetic divergence. ... *Speciation might often owe more to
ephemeral and essentially arbitrary events that cause reproductive
isolation than to the gradual and regular tug of natural selection
that draws a species into a new niche.* (Our emphasis)

> Chris Venditti (Evolutionary Biologist, University of Reading, UK) and
> Mark Pagel (Microbiologist, University of Reading, UK) 2009

In summary: We have seen how several of the recent discoveries in
biology that our book recounts lead some biologists to explicit non-
Darwinian conclusions. Samir Okasha pushes them aside saying

(correctly) that *'they simply concern aspects of biology about which traditional neo-Darwinism didn't have much to say'*. But our point about these biological mechanisms is not that the neo-Darwinists don't attend to them; but rather the marginalization of TNS that they suggest. It seems that most of the action may well be in a different part of town.

Replies to critiques from biologists

What follows are brief replies to criticisms that some of our biologist reviewers have made and that we think are radically wrong-headed; they don't exhaust the list, but they are typical.

Nothing new

A frequent critique we have received is that all the non-selectionist factors and processes summarized in Part One of our book have been known to evolutionary biologists for a long time and are all perfectly compatible with the Theory of Natural Selection (TNS). This is wrong on two counts: First, because we have based that part of our book mostly on articles published in the last five years in specialized biology journals, and (rightly) presented as innovative by their authors; Second, because it is very hard to reconcile these discoveries with TNS, as several authors say explicitly (see the quotes above and more in our book) and almost all of them at least implicitly.

In particular, our critics say that the existence of internal constraints on possible phenotypic variation is obvious and has been acknowledged to be so for decades, indeed by Darwin himself. We have doubts about this. Although we are no experts of Darwin's publications, those who are say what follows:

> There can be no direction imposed on evolution by factors internal to the organisms, because the variation upon which selection acts is random in the sense that it is composed of many

different and apparently purposeless modifications of structure. The environment determines which shall live and reproduce, and which shall die, thus defining the direction in which the population evolves.'

Bowler, 2003, pp. 10–11

One more qualified quote, by the bio-physicist and bio-mathematician Stuart Kauffman, a pioneer in the search of physical and self-organizational components of biological structures and evolution, a scientist highly regarded by Richard Lewontin and the late Stephen Jay Gould (see Chapter 5):

> A curious, logically unnecessary, but influential feature of Darwin's thinking was that the variation within one species which paved the way for emergence of well-marked varieties constituting two species *was an indefinite range*. The idea that variations could occur *in virtually any direction*, an idea which dominates in Darwin's work despite attention to correlations among traits under selection, has had important conceptual consequence. It follows that *selection alone can discriminate which new variants will be found in later generations*. Here is one root of our current idea that selection is the sole source of order in the biological world'. (Emphasis ours.)

Kauffman 1993, p. 6

Two wrong analogies

We like good analogies, but there are limits. The ones we're about to quote seem to us beyond the pale; the kind of far-fetched arguments that responsible scientists should avoid.

Thus, the authors argue, there cannot be a universal theory of natural selection, for no general relationship of phenotype to fitness can be specified. But the same might be said of many other research programs. For example, the effect of an enzyme is highly context-dependent, so Fodor and Piattelli-Palmarini presumably would not expect any successful theory in biochemistry.

Douglas Futuyma, 2010, p. 692

The net effect of an enzyme is to catalyze (that is drastically accelerate) a chemical reaction. This action depends on factors such as temperature, acidity, concentration of the substrate and of other chemical participants (co-enzymes, inhibitors). The influence of each of these factors is well understood and separable in principle. Indeed there are general laws of enzymology, such as the Michaelis-Menten equation of enzyme kinetics. These processes take place at one well specified level, that of molecular reactions, where the panorama is totally different from the highly composite one of the genotype to phenotype relation, where we have multiple levels (from Angstroms to yards), and multiple kinds of dynamics. In our book we summarize more than a dozen of these processes; the likelihood of unifying all of them under one theory is negligible. The analogy with enzymology is, therefore, totally fallacious.

The next one is due to Jerry Coyne:

> Clearly, F&P are confusing our ability to understand *how* a process operates with *whether* it operates. It's like saying that because we don't understand how gravity works, things don't fall.' [...] Our inability to understand all the details [of natural selection] is hardly a reason to claim that natural selection doesn't work.

Coyne, 2010

We are not only scientific realists, but scientific hyper-realists. Nothing like the above ever crossed our minds. We will go back to the analogy with the law of gravity in a moment, in our reply to Elliott Sober. Let's concentrate here on just one point. It's one thing to lament our failure to understand some or other natural process which we nevertheless have good reason to believe occurs. It's quite another to offer principled reasons why some or other theory of such a process isn't viable. Our book is concerned with the latter in the case of the theory of natural selection. Coyne needs to rebut these arguments. He doesn't.

We never said that NS does not operate in the wild because it's so hard for us to understand how it works. We say that general explanations based on natural selection are necessarily based on correlations (between the presence of a trait and greater reproductive potential), not causes. Detailed, very heterogeneous explanations of the selection for individual traits, in individual species, in their particular environments, can sometimes reveal causal factors. There is a radical difference, on which we insist in our book and in this update. The analogy with gravity is untenable. Gravity is the cause of the falling of bodies, not a correlation.

Merging evolution and Natural Selection

In his review, and in his recent book, Coyne regularly fails to distinguish arguments about evolution and arguments about natural selection. For example, Coyne and Dawkins both discuss at length the circuitous and devious geometry of the laryngeal nerve in mammals, which connects organs only a few inches apart, but runs from the head to the heart, looping around the aorta and then doubling back up to the neck (Coyne points out that, in the giraffe, this detour involves about fifteen feet of superfluous nerve). Then follows an account of how this oddity occurred via progressive transformations from older species of the anatomy of the organs, something we have no reason to question. Dawkins and Coyne take such cases to argue against evolution by 'intelligent design', and so they do. They are, however, thoroughly irrelevant to the issues that our book is concerned with, which is whether the mechanism of evolution is Natural Selection. But then, these data and arguments in favour of the evolutionary descent of species are transmuted into data and arguments in favour of the theory of natural selection. Questioning TNS is considered identical with questioning evolution as such. This conflation leads Coyne to say:

Their claim to have nullified 150 years of science, and one of humanity's proudest intellectual achievements, with some verbal legerdemain, is not only breathtakingly arrogant but willfully ignorant of modern biology.

Enraged at having failed to hit the target he intended, Coyne proceeds to loose his shafts at a venture.

We repeat: We have no doubts about the reality of evolution, or, more specifically, about the descent and radiation of species from preexisting ancestors; and we entirely accept that topological and functional transformations of internal organs offer persuasive evidence in its favour. What we seriously doubt is the power of natural selection to explain how it happens.

The argument from the success of artificial selection

Here's another argument of Coyne's:

> If there really were so many constraints on selection, and if development really were so complex and tightly interconnected that organisms could not respond to natural selection, then why would artificial selection be so effective at changing animals and plants?

First of all, we do not say that 'organisms could not respond to natural selection'. What we say is that there are innumerable ways of responding, depending on the phenotype, the species and the environment, defying a unitary theory. Moreover, to the best of our knowledge, artificial selection has never managed to produce new species, something that natural selection is supposed to have done many times. So, even artificial selection is effective only up to a point. Numerous sub-species have been obtained, by means of repeated selective cross-breeding, aiming at specific phenotypes (better wool, more milk, stronger muscles etc.). In our book (p. 62 and p. 210, n. 2) we stress that these desired traits were invariably accompanied by a number of others (curly tails, floppy ears, piebald colour etc.). These other traits are free-riders that were obviously not selected for. The lesson here is that, in cases of artificial selection, it's straightforward to decide which trait was selected for and which one came fortuitously, because we can ask the human agents involved, or make an educated guess. The burden of our book is that, on the one hand, the distinction between traits that are selected for is essential to distinguishing causes of fitness from free-riders; and, on the other hand, this distinction can't be drawn in cases where there *isn't* a breeder (including, in particular, cases of selection in the wild).

Missing heritability

Coyne makes the following accusations:

> Beyond distorting the scientific literature, F&P make a number of claims that are simply silly. I mention just one: 'The textbook cases of Mendelian inheritance, in spite of their great historical and didactic importance, are more the exception than the rule.' This came as a surprise to me. In fact, cases of Mendelian inheritance (the random assortment of parental genes into sperm and eggs) are the rule; if they weren't, genetic counseling would be useless. Statements like this typify the authors' attitude toward science throughout their book: they seize on some new wrinkle in the scientific literature, like a rare gene that doesn't behave according to Mendel's rules, and interpret it as a revolution that nullifies all of mainstream biology. This lack of grounding is often seen in work by science journalists who make their living touting 'revolutionary' new findings, but it is inexcusable in a supposedly serious book written by academics.

We are not surprised that this came as a surprise to Coyne. Indeed genetic counsel is often (not always, but often) useless, for instance, when well characterized frequent mutations in over twenty genes explain just three per cent or five per cent of genetic risk. The case of the 'missing heritability of complex diseases' is not a 'wrinkle', as

Coyne would have us believe. Witness the manifesto by this title published in *Nature* (8 October, 2009, Vol. 461, pp. 747–53) by twenty-seven leading human geneticists lamenting the situation, and the following summary by one of the authors, David Goldstein (Richard and Pat Johnson, Distinguished University Professor, Director, Center for Human Genome Variation, Duke University) in the *New England Journal of Medicine* on 23 April, 2009:

> Twenty gene variants account for three per cent in the variation of risk susceptibility to type 2 diabetes ... If common variants are responsible for most genetic components of type 2 diabetes, height, and similar traits, then genetics will provide relatively little guidance about the biology of these conditions, because most genes are 'height genes' or 'type 2 diabetes genes ... News are as bleak as they could be.

These are not the irresponsible scientific journalists to whom Coyne compares us. A quote will say it all. Another of those authors, Leonard Kruglyak (Professor of Ecology and Evolutionary Biology at Princeton University) in *Nature*, Vol. 456, 6 November, 2008, p. 21 says:

> It's a possibility that there's something we just don't fundamentally understand, that it's so different from what we're thinking about that we're not thinking about it yet.

Kruglyak refers to the genotype-phenotype relation for complex diseases, but the same can be said, we think, for complex traits more generally. We suggest that Coyne absorbs these facts, stops pontificating and pays attention, not to us, but to these colleagues of his.
Coyne concludes:

> In the end, F&P's contrarian efforts are all belied by the world of Richard Dawkins – the flourishing field of modern evolutionary biology, where *natural selection remains the only explanation* for the wondrous adaptive complexity of organisms.

Catching phenotypes

We conclude our replies concerning biology with a critique voiced both by Douglas Futuyma and Jerry Coyne:

> The ludicrous analogy with which Fodor and Piattelli-Palmarini end: 'organisms "catch" their phenotypes from their ecologies in something like the way that they catch their colds from their ecologies'.
>
> Futuyma

> After much demurring, they float the idea that 'organisms "catch" their phenotypes from their ecologies in something like the way that they catch their colds from their ecologies'. Although this 'explanation' links evolution to ecology, it's completely meaningless. How did ancestral whales catch their flukes and flippers from the water? How did ancestral birds catch their wings from the air? F&P don't say.
>
> Coyne

Actually, we don't think that whales catch their flukes from the water. This discussion is, of course, awash in metaphors on both sides, and the thing about metaphors is that if you don't treat them with a dollop of subtlety, they are likely to bite you. Darwin's metaphor is: 'Natural selection is like breeding'. We think it invites failures to notice the difference between breeding-for (which is intensional) and selection (which is not). Our metaphor is: 'the processes that mediate coming down with a phenotypic trait are like the ones that mediate coming down with a cold'; the point is that both depend on massive dynamic interactions between a host's endogenous properties and properties in its environment; and quite likely the details of such interactions are highly idiosyncratic from case to case. That's why nobody in his right mind thinks there could be a general theory of catching diseases. Why, then do biologists think there could be a general theory of the evolution of phenotypes?

Replies to critiques of the conceptual situation
Explanations and definitions

The crucial sentence in Peter Godfrey-Smith's review of *What Darwin Got Wrong* in the *London Review of Books* is:

> If one [but not the other of two linked traits] is causing increased reproductive success, *it is* [sic] being selected for, in the sense that matters to evolutionary theory.

A number of other reviewers have made much the same suggestion, but it won't do. The theory of natural selection claims that a trait's having been selected for causing reproductive success *explains* why a creature has it. But then it can't also claim that 'in the sense that matters' 'a trait was selected for' *means* that it is a cause of reproductive success. For, if it did mean that, then the theory of natural selection would reduce to *a trait's being a cause of reproductive success explains its being a cause of reproductive success* which explains nothing (and isn't true).

This is all old news; because John's being a bachelor *is* his being an unmarried man, John's being a bachelor doesn't *explain* his being an unmarried man. Psychologists who hoped to defend the 'law of effect' by saying that it is *true by definition*, that reinforcement alters response strength, made much the same mistake that Godfrey-Smith does.

Likewise, Elliott Sober says,

> The distinction between selection-for and 'free riding' is nothing other than the distinction between cause and correlations.

Later on he says:

> There is selection for trait T in a population if and only if trait T causes organisms to have reproductive success in the population.

This, he claims, is a *definition* of 'selection-for': it's true by definition that the trait that is a cause of increased fitness is selected-for but the other is not. However, as we just saw, that can't be right. The very heart of TNS is the thesis that, in the paradigm cases, traits are selected-for *because* they are causes of fitness; that is, differences of their effects on fitness *explain* why some traits are selected-for and others aren't. But if that's so, then the connection between being selected-for and being a cause of fitness can't be *definitional*. The dialectics here precisely parallel arguments that philosophers of mind offered in the fifties against the claim that, in paradigm cases, the relation between behavior and mental states is 'criterial' (in effect, definitional). If it's *conceptually necessary* that you raise your arm when you want to, then the cause of your raising your arm can't be your wanting to raise it. It took fifty years for philosophy to get over this. Must we now have it yet again? Something really *is* seriously wrong with the theory of natural selection, and stipulating that it is true by definition won't fix it.

The intensionality of selection-for

Elliott Sober has what seems to us to be a distorted view of the present polemical situation.

> Fodor and Piatelli-Palmarini really do maintain that there cannot
> be natural selection for one but not the other of two traits that
> are locally coextensive. However, in Fodor and Sober (2010) Fodor
> denies that the book says this.

Sober says, that the book says, that there can't be a *causal theory* of 'selection-for'. But the book doesn't say what Sober says it does. The book says that the Theory of Natural Selection can't provide an account of natural selection (because it's a causal theory and selecting-for is an intensional relation). So the book proposes a dilemma: either there is no such thing as natural selection, or, if there is, the Theory of Natural Selection misdescribes it.

Can linked properties be distinct in causal role?

Here's what Ned Block and Philip Kitcher think is one of our two main errors.

> Their specific charge is that, with respect to correlated traits in organisms – traits that come packaged together – there is no fact of the matter about which of the correlated traits causes increased reproductive success.

BK then speculate that we endorse the 'very ambitious' claim that when traits are correlated, there can be no fact of the matter about which trait causes what. But, of course, we don't believe, still less make, either of these claims. In fact, we think that it's preposterous on the face of it. Indeed, if the causal powers of linked traits can't be distinguished, it would not be an argument against the Theory of Natural Selection that it fails to distinguish them. We therefore spent a whole chapter (Ch. 7) discussing a number of ways in which the causal roles of confounded variables can be, and routinely are, assessed. The most obvious of these is J. S. Mill's 'method of differences': run an experiment in which one but not the other of the putative causes is suppressed. If you still get the effect, then it must be the variable you *didn't* suppress that's doing the causing. People (scientists very definitely included) do this sort of thing all the time, and with great success. All this is familiar from Phil. 101. Do Block and Kitcher really believe that, old and battle-weary as we are, we could have written a book that gets that wrong? The question whether there is a fact of the matter about which variable is the cause, or about whether this fact of the matter is epistemically accessible, really must not be confused with whether Natural Selection, as Darwin understood it, is able to distinguish causes from their local confounds. For reasons the book details, we think it can't. To repeat: One can work out what caused what in all sorts of ways: use Mill's method; or take the system of causes and effects apart and find out what mechanisms operate inside it; or ask the guy who built it (if somebody did) how it works ... and on and on and on. But Natural Selection can't do any of these things. It can't look inside, and it can't run experiments, and it can't contrive theories, and it can't consult the intentions of the builder. All natural selection can do is recognize *correlations* between phenotypic traits and fitness. And that doesn't help because, by assumption, if either of the confounded traits is correlated with fitness, so too is the other, and to the same extent.

Samir Okasha, in his review, commits much the same misreading of our book. He accuses us of denying the distinction between causes of fitness and free-riders. But our view is *neither* that it is impossible to deconfound causes of fitness from free-riders *nor* that there is no such distinction. What we do think (and what we think our book shows) is that Darwin's theory can't, even in principle, specify a mechanism by which selection could reliably distinguish causes of fitness from correlates of causes of fitness. To a first approximation, this is because TNS recognizes only exogenous variables as selectors, and the only (relevant) fact to which such variables are sensitive, according to TNS, is the strength of the correlations between phenotypic changes and changes of fitness. And, of course, correlation doesn't imply causation. Indeed it *patently* doesn't imply causation when the correlation in question is identical for both of the candidate causes; as it is by assumption, in the case where phenotypic traits are linked.

To repeat: It is beside the point that scientists in the laboratory often can deconfound linked causes; scientists have minds and the process of evolution does not. Indeed, it is the prima facie connection between intensional states and mental states that makes the intensionality of 'select for' a problem for naturalising TNS; a point in respect of which *What Darwin Got Wrong* is vehement.

For a while it bothered us that many of our critics should have so blatantly misread what we wrote. But we have a theory: It's that the neo-Darwinian community is so blindly committed to TNS that they allow themselves to reason as follows (implicitly, to be sure):

(1) This book says that TNS can't distinguish causes of fitness from correlates of causes of fitness. But, it goes without saying that (2) TNS is certainly true and everybody knows that it is. So (3) if the authors claim that TNS can't distinguish causes from correlates, that

must be because they think that there is no such distinction. So (4), I shall write a review accusing them of thinking that. But if that is indeed how our critics are reasoning, we protest that it's more than a tad question-begging.

Laws of evolution

A short summary of the second half of the book might go like this: TNS needs selection-for to be intensional, but offers no suggestion of how it could be. But, as we remarked above, if there are laws of evolution (nomologically necessary empirical generalisations to which evolutionary processes conform) it might be from those that the intensionality of select-for derives. So it matters to the present question whether there are such laws. The bad news, according to *What Darwin Got Wrong*, is that there aren't. This is, indeed, one of the cases in which WDGW agrees with what we take to be the consensus view among biologists. Nobody doubts that evolution is law-governed; after all, the laws of physics apply to everything. The present issue is whether there are *biological* laws of evolution; that is, laws of evolution that are defined over *biological* kinds (such as, for example, laws about evolution defined over ecological properties *so described* and their effects on fitness *so described*). Missing this point has lead to all sorts of confusion including, notably, the suggestion that if there are no laws of evolution, determinism and/or mechanism are ipso facto undermined.

Well, Elliott Sober thinks we're wrong about that. Actually what he says is not that there are such laws, but that we haven't shown that there aren't. And indeed we haven't. Since the issue is entirely empirical, there's no question of demonstrative arguments on either side. There are, however, straws in the wind, and we think they're blowing our way.

Here are two reasons for doubting that there are laws of evolution. The first is that there seem to be no examples of such laws. That is easily explained on the assumption that, in fact, there are no such laws. The second is that, if there were laws of evolution, they would have to be horrendously complicated. A long tradition of modelling evolution has indentified at least the following factors, among others: effective population size, density-dependent selection, drift with or without selection, migration, gene flow and horizontal transmission, the diffusion of neutral mutations, mutational bias, biased gene conversion, differentials in fertility, sexual selection, variable sex ratios, the overlap of fertile generations, the fixation of deleterious alleles, phenotypic plasticity, and various kinds of epistasis (gene-gene interactions). Sober says (rightly) that complexity isn't, in and of itself, an argument against the putative laws. But the kind of complexity that laws of evolution would require is, we think, without precedent in the other sciences. First of all, laws of evolution would have to take into consideration interactions at vastly heterogeneous levels: molecule to molecule, gene to gene, gene to cell, cell to cell, developmental module to developmental module, tissue to tissue, organism to organisms of the same species, organism to organisms of different species, and all these to the local ecology. The heterogeneity concerns both sheer size (from Angstroms to miles) and the conceptualisation of the relevant kinds. His failure to understand this is part and parcel of Sober's mishandling of one of his own examples:

> The gravitational force now acting on the earth depends on the mass of the sun, the moon, and of everything else. It does not follow that there are no laws of gravity, only that the laws need to have numerous placeholders ... The fact that an effect has numerous complexly interacting causes does not show that there are no laws about this complex cause/effect relation.

Well, of course there are laws of gravity; principally that the gravitational force between objects varies directly with their total mass and inversely with the square of their distance. Notice, however, that this law is quite simple; in particular, it has no 'place holders' for the sun, the moon, the Earth or anything else except the masses and distances of the objects involved. That's why the law of gravity would be unaffected even if there weren't the sun, the moon, or the earth.

What goes on when explanations appeal to laws is something like this: there are variables for relevant properties of things that fall under the laws; and there are specifications of the 'initial conditions' in some domain to which the laws apply. Neither the moon nor its mass gets mentioned by the laws of gravity; but both *do* get mentioned in specifying the conditions that obtain when the theory of gravity is used to predict the gravitational force between, for example, the moon and the earth. In consequence, the laws of gravity have very many fewer 'place holders' than there are things in the universe to which they apply. We won't argue for this view; but please take our word for it that a lot depends on getting it straight.

So now the question arises whether this picture is plausible for the (putative) evolutionary laws of trait fixation. We think it pretty clearly isn't; *not,* however, because there are very many creatures to which the laws would have to apply, and very many environmental features with which such creatures may interact. Rather, it's because of the awesome *heterogeneity* of levels and kinds we have mentioned, and of the ways in which interactions of creatures with their environment depend on what kind of creature it is and what kind of environment it is interacting with. As we saw two paragraphs back, laws don't need place-holders for each thing that falls under them, but they do need placeholders for each *kind* of thing that falls under them.

To make the point slightly differently, there are typically many kinds of creatures that can share an environment, and many kinds of environments that creatures can share. (We're told that more than ten thousand species share Central Park.) That being so, the putative laws that determine fitness as a function of such interactions would have to be complicated in precisely the way that the laws of gravity are not: They would need 'place holders' for each of the *kind* of creatures that they apply to and for each *kind* of environment that the creatures can interact with. And, to repeat, though the number of things a law applies to doesn't determine how many placeholders it needs, how many *kinds* of things it applies to does. Given all that, could there be such laws about how creature/environment interactions determine fitness? In principle, sure there could. But are there such laws? We think the probability is asymptotically close to nil. The kind of complexity that *does* tell against a putative law is the kind that proliferates kinds beyond necessity.

There are other things Sober's review says that we think are wrong; for example, we think it's wrong about whether truths about individual events support counterfactuals (except for the dreary counterfactual that if exactly the same thing were to happen again, all else being equal, exactly the same effects would ensue). But, for present purposes, we're content to leave it here.

TNS versus sufficient reason

David Papineau, in his review says:

> If Fodor and Piattelli-Palmarini are right, polar bears don't have white fur because it confers advantages in the Arctic; we don't have eyes because they help us to see; and in general there is no tendency for natural selection to preserve adaptive traits.

Could we really be denying that the reason polar bears are white is that being white hides them in the snow? No. Part of the story about why polar bears are white is surely that there were many causal chains in which white polar bears got missed by their predators (and/or were able to sneak up on their prey) more regularly than polar bears that were less white. On our view, tracing such causal chains is what natural history does for a living. But a theory of Fs doesn't consist of an enumeration of causal chains in which Fs are involved. *A theory of Fs is an account of what Fs have in common as such.* Accordingly, a theory of trait evolution is an account of what instances of trait evolution have in common as such. (Notice, in passing, that 'as such' is intensional). So what does TNS say about what instances of trait evolution have in common as such? What, for example, does it say about what the evolution of four chambered hearts in mammals, and of long necks in giraffes, and of web spinning in spiders and of bipedal gait in us have in common *qua* instances of trait evolution?

Just this: *In every such case there has to be something about the creatures (or about their ecology, or both) such that those of the creatures that were F flourished more than otherwise similar creatures that were not F.* Well of course there has to be. That follows just from the 'principle of sufficient reason' according to which if something is F, there must be something that caused it to be F; and of course, whatever the 'something' is, it has to be either internal to the organism or external to the organism. There's no other place that it could be. On our view *there is no theory of evolution.* All there is, is natural history.

Speaking of the adaptive function of the eye (as Papineau urges us to do) a species of jellyfish (the cubozoan jellyfish, *Tridpedalia cystophora* discovered in the waters near Puerto Rico) has twenty-four globular eyes in six groups of four (called rhopalia), very similar to our vertebrate eyes, but no brain to collect the images, no optic nerve, and the lenses can only form images behind the retina. No adaptive explanation is in sight, though the genetic and developmental mechanisms responsible for this feast of structure without function are well understood.

On mathematical models

Samir Okasha and other reviewers hope to vindicate TNS by appealing to the 'paradigm' (sic) explanatory power of mathematical models of natural selection. We are fully aware of the long and illustrious tradition of mathematical theory of natural selection and, more generally, of evolution; from the Hardy-Weinberg law of equilibrium between allele frequencies (1908) to the works of Ronald Fisher, J. B. S Haldane and Sewall Wright (1924–37) to George R. Price's theorem (1970, 1972) all the way to the present day (for thorough expositions see Provine 1971/2001 and Rice 2004). However, as William Provine, a leading historian of mathematical evolutionary theories, says:

> They [Fisher, Haldane, Wright, Hogben, Chetverikov and other
> mathematical modelists] all disagreed, often intensely, with each

other about actual processes of evolution in nature, *even when their models were mathematically equivalent*. (Our emphasis)

William B. Provine, 1988, p. 56

Several critiques of the plausibility of many such models have been raised by qualified biologists including, to name a few, Carl Woese, Andre'Ariew and Richard Lewontin, Richard Michod and even Massimo Pigliucci, who is by no means in sympathy with our view of TNS. In particular, Carl Woese, in a recent interview with Marc Buchanan for the *New Scientist*, says:

Biology built up a facade of mathematics around the juxtaposition of Mendelian genetics with Darwinism, and as a result it neglected to study the most important problem in science – the nature of the evolutionary process.'

Buchanan, 2010

And it is again beside the point that scientists are quite often successful in constructing models of such phenomena as the evolution of sex ratios in a population; or of how actual foraging strategies approximate ideal foraging strategies, etc. The point is that such models aren't causal explanations; they don't do – they don't even purport to do – what so many proponents of TNS claim that it does: explicate the causal mechanism of evolution. The most strenuous defenders of the modern synthesis state explicitly that, although causal inference is desirable, mathematically, all that is required is correlation. In general, mathematical models can only be as good as the idealisations on which they are based. In the words of a leading expert and author of a comprehensive technical treatise:

It is in the nature of model building that our models often hinge on assumptions that we know are not exactly true. What is interesting about [two such] assumptions – monomorphic populations in which variant strategies appear one at a time and populations that respond quickly to environmental changes – is that they are contradictory. A population cannot quickly evolve to a new equilibrium unless it has a substantial amount of heritable variation. If evolution always had to wait for a new variant to arise by mutation, it would be a very slow process, especially if each new mutation differed from the previous state by only a small amount. Thus, when one of these assumptions is a good approximation, the other one ceases to be.

Sean H. Rice, 2004, p. 289

Mathematical model building can make explicit the consequences of certain idealisations, but it doesn't even purport to *reveal* the causal mechanisms that sustain the phenomena; whereas our worry about TNS is that no causal mechanism could do what it claims that the process of selection-for does.

Conclusion

We continue to believe that there's a lot that Darwin Got Wrong. We continue to believe that the issues implied by the externalism of his account of selection, and by his failure to notice the intensionality of selection-for, are in need of thorough and careful consideration. Thus far, the critical responses to our attempts have not been edifying; mostly a howl of reflexive Darwinism, with very little attention paid either to the structure of the arguments or to their repercussions. But we're told that hope springs eternal. Our hope, at a minimum, is to have cleared the ground for calmer and much more responsible polemics. We still believe in the possibility of a rational, interdisciplinary discussion of the empirical warrant and the conceptual architecture of TNS. But we must admit that we don't believe in it now as much as we did a year ago.

APPENDIX

'That' kind of Darwinian in psychology and the philosophy of mind

The deep influence of Darwinian theory on behaviouristic psychology has been analysed in detail in Chapter 1. In this appendix, we have assembled representative quotes from the work of evolutionary epistemologists, evolutionary psychologists and evolutionary philosophers of mind, extending all the way up to 2009. In these fields, *pace* our friends in 'wet' biology, there were, and there are, *that* kind of Darwinian, which is to say, unabashed adaptationists, and they are quite influential. We start with a brief summary of the history of the influence of Darwinism on theories of mind and behaviour (for a detailed reconstruction, see Richards, 1987).

From Boltzmann to W. V. Quine

In a popular lecture delivered in Vienna in 1900, the physicist Ludwig Boltzmann, one the fathers of statistical mechanics and the kinetic theory of gases, declared that the nineteenth century would be remembered as the Century of Darwin, then stated:

> In my view all salvation for philosophy may be expected to come from Darwin's theory. ... What then will be the position of the so-called laws of thought in logic? Well, in the light of Darwin's theory they will be nothing else but inherited habits of thought. ... One can call these laws of thought a priori because through

many thousands of years of our species' experience they have become innate to the individual, but it seems to be no more than a logical howler of Kant's to infer their infallibility in all cases. According to Darwin's theory this howler is perfectly explicable. Only what is certain has become inheritable; what was incorrect has been dropped. In this way these laws of thought acquired such a semblance of infallibility that even experience was believed to be answerable to their judgment.

Boltzmann then proceeds to say that theories and deductions are not first true and then, as a consequence, useful, but rather they were first useful and then, as a consequence, considered true (Boltzmann, 1904, p. 193 and *passim*).

The epistemology of William James rested on Darwinian principles. The mind comes already outfitted with fixed sensory and emotional responses, instinctive reactions, and basic rational abilities; these constitute our evolutionary legacy. But the acquisition of new ideas is also Darwinian; spontaneous hypotheses, guesses and notions erupt in our pedestrian and scientific encounters with the world; those that survive the pitiless force of reality live for another day.

Richards, 1987, p. 440

Logicality in regard to practical matters (if this be understood, not in the old sense, but as consisting in a wise union of security with fruitfulness of reasoning) is the most useful quality an animal can possess, and might, therefore, result from the action of natural selection; but outside of these it is probably of more advantage to the animal to have his mind filled with pleasing and encouraging visions, independently of their truth; and thus, upon unpractical subjects, natural selection might occasion a fallacious tendency of thought.

Peirce, 1877, p. 2

In an open course at Columbia University on 'Charles Darwin and His Influence on Science' in the Winter and Spring of 1909, John Dewey states that the influence of Darwinism on philosophy 'resided in its having conquered the phenomena of life for the principle of transition, and thereby freed the new logic for application to mind and morals and life' (reprinted in Dewey, 1910, p. 9).

Several decades later, the prominent American logician and philosopher Willard Van Orman Quine wrote:

> Why does our innate subjective spacing of qualities accord so well with the functionally relevant groupings in nature as to make our inductions tend to come out right? Why should our subjective spacing of qualities have a special purchase on nature and a lien on the future? There is some encouragement in Darwin. If people's innate spacing of qualities is a gene-linked trait, then the spacing that has made for the most successful inductions will have tended to predominate through natural selection. Creatures inveterately wrong in their inductions have a pathetic but praiseworthy tendency to die before reproducing their kind.
>
> Quine, 1969, p. 13

Quine was too subtle a philosopher to be fully satisfied by this explanation. He acknowledged that there is a circularity in it. One instance of induction (the Darwinian theory) in one special science (evolutionary biology) cannot offer a firm foundation for the success of induction in general. But in his lifetime endeavour to naturalize epistemology and to make of philosophy a science among others, he thought that this kind of circularity was not, after all, too disturbing.

At a lower level of sophistication were the American psychologist and social scientist Donald T. Campbell and the Austrian–British epistemologist Karl R. Popper. Campbell is credited with having coined the expression 'evolutionary epistemology', which was later, and more successfully, publicized by Popper. In Campbell's definition evolutionary epistemology is 'an epistemology taking cognizance of and compatible with man's status as a product of biological and social evolution' (Campbell, 1974, p. 413). Presenting evolution itself as a cognitive process of increasing knowledge, Campbell writes: 'the natural-selection paradigm for such knowledge increments can be generalized to other epistemic activities such as learning, thought and science' (ibid.). In the wake of Boltzmann's (mis-)treatment of Kant, Campbell has characteristically also published a paper section entitled: 'Kant's categories of perception and thought as evolutionary products' (in Radnitzky *et al.*, 1987). (A complete bibliography of evolutionary epistemology up to 1985 is to be found in Campbell *et al.*, 1987).

The progress of scientific ideas and theories was, according to Karl R. Popper, a process of Darwinian selection, whereby we 'get rid of a badly fitting theory before the adoption of that theory makes us unfit to survive' (Popper, 1975, p. 78). Through this Darwinian process 'we let our [refuted] theories die in our stead' (ibid.). Although Popper had famously claimed that 'Darwinism is not a testable scientific theory but a metaphysical research programme' (Popper 1976, p. 151), he happily recanted and stated: 'And yet, the theory is invaluable. I do not see how, without it, our knowledge could have grown as it has done since Darwin' (ibid., pp. 171–72).

Evolutionary psychology (1988–2009)

It would be impossible, for reasons of space, to extract even an adequate selection of citations from the works of Richard Dawkins, Daniel Dennett and Steven Pinker. We can only refer the reader to their many books and articles. The following quotes are extracted from a huge literature on evolutionary psychology, and they are, we think, representative of the guiding ideas of other modern neo-Darwinian psychologists and philosophers. The 1028 pages of the *Handbook of Evolutionary Psychology* (Buss, 2005) lists as 'luminaries' Leda Cosmides, John Tooby, Don Symons, Steve Pinker, Martin Daly, Margo Wilson and Helena Cronin. We have here quotes from some of them, and from some others as well.

The 1988 book *Homicide* by Martin Daly and the late Margo Wilson, Professors of Psychology at McMaster University in Hamilton, Ontario, Canada, made quite a sensation. Inspired, as they say, by the theory of Darwinian evolutionary processes, they extracted evidence for a deep link between genetic relatedness (or lack thereof) and abuse and violence in step-families. They coined the term 'Cinderella effect', which was destined to become a standard in this literature:

> We are both psychologists by training, and we are inspired by the potential of selection thinking as metatheory for psychology. The entire social scientific enterprise is concerned with the characterization of human nature. How could Darwin's more encompassing theory of organismic nature – so heuristic in so many areas of the life sciences and unquestionably correct in its basics – how could it not be relevant to the task? The development of an evolutionary psychology is inevitable and to be welcomed. It will use selection thinking to generate testable hypotheses about motives and emotions and cognition and child development. It will link psychological processes both to their behavioural outcomes and to the selective pressures that have shaped them. This book is an effort in these directions.
>
> Daly and Wilson, 1988, p. 9

> Given the ubiquity of abused stepchildren in folklore and the pervasive negative stereotyping of stepparents, any child-abuse researcher might have wondered whether steprelationship is a genuine risk factor, but in fact, those whose imaginations were uniformed by Darwinism never thought to ask. We conducted the first comparison of abuse rates in stepfamilies versus intact birth families, and the difference turned out to be large.
>
> Daly and Wilson, 2008, p. 383 (see also Wilson et al., 1980)

Robert Wright, Schwartz Senior Fellow at the New America Foundation, author of *The Moral Animal: Evolutionary Psychology and*

Everyday Life, named by the *New York Times Book Review* as one of the 12 best books of 1994 and published in 12 languages:

> The thousands and thousands of genes that influence behaviour – genes that build the brain and govern neurotransmitters and other hormones, thus defining our 'mental organs' – are here for a reason. And the reason is that they goaded our ancestors into getting their genes into the next generation. If the theory of natural selection is correct, then essentially everything about the human mind should be intelligible in these terms ... Natural selection has now been shown to plausibly account for so much about life in general and the human mind in particular that I have little doubt that it can account for the rest.
>
> Wright, 1994, p. 28

Philosopher of biology, psychology, mind and language, Ruth Garrett Millikan, at the University of Connecticut and recipient of the Jean Nicod Prize in 2002:

> The mechanisms at work in stabilizing the functions of public language devices are, in certain crucial respects, much like those at work in biological evolution under natural selection. Especially obvious is the similarity between stabilization of the functions of various public language forms through social selection processes and stabilization of the functions of animal signs, such as mating displays, danger signals, territory markers, bee dances and so forth, through genetic selection processes.
>
> Millikan, 2001, p. 159

In sum, if we look at the whole human person in the light of our history of evolution by natural selection, minding the continuities between humans and other animals, it appears that all levels of purpose have their origin in adaptation by some form of selection. In this sense all purposes are 'natural purposes'. Even though there are, of course, many important differences among these kinds of purposes, there is an univocal sense of 'purposes' in which they are all exactly the same.

Millikan, 2004, p. 7

Mental representations couldn't represent their own purposes unless they had purposes to represent, and these purposes are derived from various levels of selection. Explicit desires and intentions are mental representations whose purposes are to help to produce what they represent. They were selected for helping to bring about the conditions they represent. They were not selected for one by one, of course, certainly not on the level of genetic evolution. Only the cognitive and conative mechanisms responsible for forming desires and intentions were designed/chosen by natural selection. They were selected for their capacity, on the basis of experience, to form representations of goals, of possible future states of affairs, which when brought about, sometimes furthered our biological interests.

Millikan, 2004, p. 13

American psychologist Leda Cosmides and (husband) psychologist and anthropologist John Tooby (Center for Evolutionary Psychology, University of California Santa Barbara):

[1] Natural selection – an invisible-hand process – is the only component of the evolutionary process that produces complex functional machinery in organisms ... [2] Natural selection builds the decision-making machinery in human minds. [3] This set of cognitive devices generates all economic behaviour. [4] Therefore, theories of economic behaviour necessarily include theories about the structure of the cognitive mechanisms that generate that behaviour. Moreover, the design features of these devices define and constitute the human universal principles that guide economic decision-making.

Cosmides and Tooby, 1994, p. 328

Evolutionary psychology is unusual and perhaps unique among theoretical orientations in psychology in the degree to which it derives principled predictions about previously unknown aspects of the species-typical psychological architectures of humans and other species ... Cosmides, Tooby, and colleagues have predicted and found a large number of patterns in human reasoning performance never before obtained experimentally, derived from the hypothesis that natural selection built specialized reasoning systems with procedures efficiently tailored to the recurrent properties of adaptive inferential problems involving cooperation.

Tooby et al., 2003, p. 860

The most basic lesson is that natural selection is the only known natural process that pushes populations of organisms thermodynamically uphill into higher degrees of functional order, or even offsets the inevitable increase in disorder that would otherwise take place. Therefore, all functional organization in undomesticated organisms that is greater than could be expected by chance (which is nearly all functional organization) is ultimately the result of the operation of natural selection and hence must be explained in terms of it (if it is to be explained at all). This is why understanding natural selection is enormously beneficial to any theoretically principled psychology. In effect, natural selection defines the design criteria to which organisms were built to conform. This is why knowledge of ancestral natural selection combined with knowledge of ancestral environments provides a principled theoretical framework for deriving predictions about the reliably developing design of the human mind. Natural selection is (a) the set of enduring, nonrandom, cause-and-effect relationships in the world that (b) interact with the reliably developing features of organisms (c) in such a way that they consistently cause some design variants to reproduce their designs more frequently than others because of their design differences.

Tooby et al., 2003, p. 362

The foundational recognition that psychological mechanisms are evolved adaptations connects evolutionary biology to psychology in the strongest possible fashion, allowing everything we know about the study of adaptations to be applied to the study of psychological mechanisms.

Tooby and Cosmides, 2005, p. 9

Selection should retain or discard alternative circuit designs from a species' neural architecture on the basis of how well the information–behaviour relationships they produce promote the propagation of the genetic bases of their designs. Those circuit designs that promote their own proliferation will be retained and spread, eventually becoming species-typical (or stably frequency-dependent); those that do not will eventually disappear from the population. The idea that the evolutionary causation of behaviour would lead to rigid, inflexible behaviour is the opposite of the truth: Evolved neural architectures are specifications of richly contingent systems for generating responses to informational inputs.

Tooby and Cosmides, 2005, p. 13

The programs comprising the human mind were designed by natural selection to solve the adaptive problems regularly faced by our hunter-gatherer ancestors – problems such as finding a mate, cooperating with others, hunting, gathering, protecting children, navigating, avoiding predators, avoiding exploitation, and so on. Knowing this allows evolutionary psychologists to approach the study of the mind like an engineer.

Tooby and Cosmides, 2005, p. 16

The fact that the brain processes information is not an accidental side effect of some metabolic process. The brain was designed by natural selection to be a computer. Therefore, if you want to describe its operation in a way that captures its evolved function, you need to think of it as composed of programs that process information.

Tooby and Cosmides, 2005, pp. 16–17

199

Philosopher of science and of psychiatry Dominic Murphy of the California Institute of Technology and philosopher and cognitive scientist Stephen Stich of Rutgers University:

> A central tenet of evolutionary psychology is that the human mind is designed to work in our ancestral, hunter-gatherer environment. Natural selection did not design it for the contemporary world. But, of course, a system may function admirably in one environment and work rather poorly in another. So it is entirely possible that the mind contains modules or other sorts of systems which were highly adaptive in the ancestral environment but which do not lead to functional behaviour in our novel modern environments.

> Murphy and Stich, 2000, pp. 18–19

> One of the morals to be drawn from ... hypotheses about depression is quite general. The environment in which selection pressures acted so as to leave us with our current mental endowment is not the one we live in now. This means that any mental mechanism producing harmful behaviour in the modern world may be fulfilling its design specifications to the letter, but in an environment it was not designed for.

> Murphy and Stich, 2000, p. 21

Psychologist and computer scientist Peter M. Todd at Indiana University (and Editor-in-Chief of the journal *Adaptive Behavior*) and psychologist and historian of probability Gerd Gigerenzer, Director of the Center for Adaptive Behavior and Cognition at the Max Planck Institute of Human Behavior, Freie Universität Berlin, Germany:

> Bounded rationality can be seen as emerging from the joint effect of two interlocking components: the internal limitations of the (human) mind, and the structure of the external environments in which the mind operates. This fit between the internal cognitive structure and the external information structure underlies

the perspective of bounded rationality as ecological rationality – making good (enough) decisions by exploiting the structure of the environment.

Todd and Gigerenzer, 2003, pp. 147–48

Standard statistical models, and standard theories of rationality, aim to be as general as possible, so they make broad and mathematically convenient assumptions about the data to which they will be applied. But the way information is structured in real-world environments often does not follow convenient simplifying assumptions ... While general statistical methods strive to ignore such factors that could limit their applicability, evolution would seize upon informative environmental dependencies like this one and exploit them with specific heuristics if they would give a decision-making organism an adaptive edge. Finding out when and how structures of information in environments can be used to good advantage by simple heuristics is thus a central component of the ecological rationality research program.

Todd and Gigerenzer, 2003, p. 151

H. Clark Barrett of the Department of Anthropology at the University of California, Los Angeles, Robert Kurzban of the Department of Psychology at the University of Pennsylvania, and Edouard Machery of the Department of History and Philosophy of Science at the University of Pittsburgh:

It is worth noting that although there are principled reasons to expect that natural selection favors specific rather than general mechanisms, this principle applies equally to function-general and function-specific systems. Even systems with very general functions – however these functions are stated – must be ones that are plausible from the adaptationist framework.

Barrett and Kurzban, 2006, p. 643

Take, for example, the problems of avoiding fitness-reducing environmental hazards such as predators, pathogens, toxins, and heights. Evolutionary psychologists, and presumably others, regard these as relatively uncontroversial adaptive problems (even many psychologists who view evolutionary psychology with skepticism believe that disgust is an adaptation to prevent ingestion of toxins and pathogens, for example). While it is not possible to predict with certainty the exact nature of solutions to these problems, it is certainly possible, and legitimate, to use the nature of these problems to generate hypotheses about the possible design features of adaptations.

Machery and Barrett, 2006, pp. 237–38

American cognitive psychologist Edward T. Cokely at the Max Planck Institute of Human Behavior, Freie Universität Berlin, and epistemologist Adam Feltz, Director of the Behavioral Philosophy Lab at Schreiner University in Kerrville, Texas:

> We suggest that philosophically relevant intuitions can be effectively theoretically characterized within an adaptive or ecological framework – i.e. a Darwinian inspired perspective on the fundamental goal-enabling nature of cognition.

Cokely and Feltz, 2009, p. 358

On the adaptive view, the goals and needs of organisms, such as finding food, securing mates, or protecting offspring, may or may not benefit from cognition that is logically coherent or philosophically invariant. For these organisms, fitness may be best served when cognition is variable, diverse, and tuned to ecological constraints and demands. An organism's success will depend on the extent to which its cognition can benefit from and exploit the fit between its internal (i.e., psychological) and external (e.g., social and physical) environments regardless of logical coherence (or lack thereof). In an uncertain and complex world such as ours, we should not expect or necessarily even want to always be governed by processes that maintain logically coherent cognition ... Indeed, it is well known that our cognitive constraints limit our ability to be 'rational,' but what is more provocative are the data indicating that cognitive constraints can facilitate adaptive judgment … Accordingly, our research program and approach are fundamentally concerned with when, how, and why adaptive psychological processes (e.g. cognitive sensitivities, affective biases, and simple heuristics) enable, track, or contradict intuitions that are produced via traditional analytical philosophy. A cornerstone of these efforts involves the investigation of the complex and varied relations between individual differences – such as personality traits – and philosophically relevant intuitions.

Cokely and Feltz, 2009, p. 358

American evolutionary and social psychologist David M. Buss, head of the Individual Differences and Evolutionary Psychology Area at the University of Texas, Austin, and editor of the vast Wiley *Handbook of Evolutionary Psychology* (2005), in a special issue of *American Psychologist* on 'Charles Darwin and Psychology':

The emergence of evolutionary psychology and related disciplines signals the fulfillment of Darwin's vision. Natural selection theory guides scientists to discover adaptations for survival. Sexual selection theory illuminates the sexual struggle, highlighting mate choice and same-sex competition adaptations. Theoretical developments since publication of *On the Origin of Species* identify important struggles unknown to Darwin, notably, within-families conflicts and conflict between the sexes. Evolutionary psychology synthesizes modern evolutionary biology and psychology to penetrate some of life's deep mysteries ….

Buss, 2009, p. 140

Although considered controversial by some, modern evolutionary psychology signals the actualization of Darwin's prediction that psychology would be based on a new foundation. His theories of natural and sexual selection identified core processes by which functional psychological mechanisms evolve – the struggle for existence and the struggle for mates. Natural and sexual selection serve important functions that characterize the best scientific theories – they guide investigators to important domains of inquiry, generate novel predictions, provide cogent explanations for known facts, and produce empirical discoveries that would not otherwise have been made.

Buss, 2009, p. 146

Evolutionary psychology has advanced beyond Darwin's vision in several ways. The first stems from theoretical developments in evolutionary theory that occurred after Darwin's day – the discovery of particulate inheritance, the modern synthesis, the theory of inclusive fitness, and the understanding of the logical implications of genic selection. The second was fashioned by the cognitive revolution – the view that psychological adaptations can be conceptualized as information-processing devices instantiated in the brain (Tooby and Cosmides, 2005). The third followed from exploring new domains, such as sexual conflict and within-family conflict, that are being illuminated by modern evolutionary theory … In 1859, Darwin provided a vision of a distant future in which psychology would be based on the new foundation. The distant future that Darwin envisioned is upon us. Modern psychologists are privileged to experience a scientific revolution that signals the realization of that vision.

Buss, 2009, pp. 146–47

We think that a fitting conclusion to this anthology is the paean addressed to evolutionary psychology by the distinguished Johnston Professor of Psychology at Harvard University, Steven Pinker, author of bestsellers, named one of *Time Magazine*'s 100 most influential people in the world in 2004, and one of *Prospect and Foreign Policy*'s 100 top public intellectuals in 2005. He was twice a finalist for the Pulitzer Prize, in 1998 and in 2003:

Evolutionary psychology is the cure for one last problem ailing traditional psychology: its student-disillusioning avoidance of the most fascinating aspects of human mental and social life. Even if evolutionary psychology had not provided psychology with standards of explanatory adequacy, it has proved its worth by opening up research in areas of human experience that have always been fascinating to reflective people, but had been absent from the psychology curriculum for decades. It is no exaggeration to say that contemporary research on topics like sex, attraction, jealousy, love, food, disgust, status, dominance, friendship, religion, art, fiction, morality, motherhood, fatherhood, sibling rivalry, and cooperation has been opened up and guided by ideas from evolutionary psychology. Even in more traditional topics in psychology, evolutionary psychology is changing the face of theories, making them into better depictions of the real people we encounter in our lives, and making the science more consonant with common sense and the wisdom of the ages.

Pinker's Foreword to Buss, 2005

NOTES

Chapter I

1. Which is roughly to say the census of heritable properties that they share. One could (and some do) argue about exactly how the notion of a phenotype should be defined, but we don't propose to become involved with this: first, because the details won't matter to the considerations we will raise here; and, second, because we think arguing about definitions is pretty generally a waste of breath.

2. This is only a rule of thumb as, presumably, some similarities among species among phenotypic properties are the consequences of evolutionary convergence. We shall assume without argument that these are occasional exceptions to the general principle, which is that phenotypic similarities are typically explained by common ancestries.

3. Similarities between OT and ET are a constantly recurring theme of Skinner's: 'Just as we point to survival to explain an unconditioned reflex, so we can point to "contingencies of reinforcement" to explain a conditioned reflex' (Skinner, 1976, p. 43). 'Evolutionary theory moved the purpose which seemed to be displayed by the human genetic endowment from antecedent design to subsequent selection by contingencies of survival. Operant theory moved the purpose which seemed to be displayed by human action from antecedent intention or plan to subsequent selection by contingencies of reinforcement ... Contingencies of reinforcement also resemble contingencies of

survival in the production of novelty ... In both natural selection and operant conditioning the appearance of 'mutations' is crucial' (Skinner, 1976, pp. 246–47).

4. What reconciles these two ways of thinking about NS is that speciation is itself a process in which the distribution of phenotypes in a population alters.

5. It is important to distinguish 'population thinking' (i.e. the metatheoretic view that a main goal of evolutionary theory is to exhibit the processes by which the relative frequency of phenotypes in a population changes over time) from the idea that populations are the 'units of natural selection' (roughly, the thesis that it is the fitness of populations, rather than the fitness of the individuals by which populations are constituted, that reliably increases under selection pressure). The importance of 'population thinking' is common ground in practically all contemporary theorizing about evolution. By contrast, claims about the unit(s) of selection are hotly disputed, especially in discussions of the evolution of altruism (cf. Sober and Wilson, 1998).

6. A creature's psychological profile is thus to be distinguished from its psychological phenotype; by stipulation, the latter includes only psychological traits that are heritable, which, by general consensus, learned traits are not.

7. The requirement that stimuli and responses both be 'publicly observable' is what makes Skinner's associationism behaviouristic.

8. They are proprietary in the sense that they hold over and above the requirement that candidate theories make empirically correct predictions about how phenotypes (behavioural profiles) vary over time.

9. If you think of operant conditioning theory and evolutionary theory as literally input–output functions, then it's both natural and harmless to think of the 'external' constraints as conditions of adequacy on how these functions are computed (hence,

indirectly, on how they are implemented). We shall talk this way from time to time.

10. For example, the requirement that the chosen theory should be 'mechanistic' and 'physicalistic' – whatever exactly that amounts to.

11. It's unsurprising that OT should have opted for the view that psychological profiles are shaped primarily by environmental variables. Skinner was a behaviourist, after all, and behaviourists don't like theories that appeal to 'unobservables' – mentalistic unobservables least of all. It is much less clear why Darwin should have taken it so much for granted that the course of evolution is primarily exogenously constrained. Unlike Skinner, Darwin wasn't under the sway of a positivist philosophy of science. This matters a lot in the present context, as a natural alternative to adaptationism might be an account of evolution in which endogenous factors play the central role. Something of the sort seems to be emerging in the 'evo-devo' movement.

12. Predictably, given their endorsement of the gradualism constraint, Darwinists generally claim that there are no 'essential' or 'defining' properties of species.

13. It isn't known whether the probable viability of large mutations is sufficiently small to preclude their contributing significantly to the observed heterogeneity of phenotypes. It isn't even known how large such large-but-viable mutations would have to be in order for them to do so. We can't imagine how, in practice, reliable estimates might be made. Sometimes the case is argued by analogy to the unlikelihood that a tornado in a junk yard would blow a Boeing 707 into being – which seems to us a thoroughly irresponsible kind of polemic, as no one has any idea what the likelihoods are in either case.

14. It is in dispute, even in paradigm cases of operant conditioning, just how smooth learning curves actually are. What appear to be smooth curves can emerge when the data are summed across subject groups. Discontinuities of the sort that insight learning

would predict can emerge when the data are analysed separately for each subject. For discussion of this and related issues, see Gallistel (2000, 2002).

15. There are, to be sure, cases of evolutionary 'regression' (e.g. fish that live in caves may lose the ability to see), but these are typically attributed to changes in the selection pressures, not to changes in the relation between selection and fitness. In a dark environment vision bestows no competitive advantages.

16. 'Efficiency' is the ratio of reinforced responses to total responses.

17. The role in evolution of statistical drift, founder effects and the like are acknowledged, but they are generally held to be significant only in relatively special circumstances (small and/or isolated populations, for example).

18. There are notorious problems about whether such claims are other than tautological: perhaps fitness just is whatever increases under selection; perhaps reinforcement just is whatever increases habit strength. We won't, however, press such objections in what follows.

19. The point in both cases is that (barring effects of random variables) the mere possibility of preferable alternatives has no effect on the actual course of either learning or evolution; if there is movement away from a local maximum, it must be fortuitous. Accordingly, if an associative network gets stuck in a local maximum, the best you can do is to 'raise the temperature': increase the frequency of random responses, some or other of which may prove more efficient than the one on which the device has thus far settled. That this seems a poor alternative to the exercise of foresight is a stumbling block for associationists in general and for connectionists in particular.

20. It is a sophisticated thing to say at cocktail parties that artificial selection is itself (just) a kind of natural selection. But it is not; not, anyhow, if Darwin was right to hold that natural selection cannot be mediated by mental causation.

21. The parenthetical caveat is because it's possible to deny that all mental causes are representational; pains, sensations and the like can be causes, and maybe some or all of these aren't kinds of mental representations. There is currently a furious debate about this in the philosophical literature, which, however, we will entirely ignore in what follows. In this book, 'mental' always means 'intentional and representational' lacking explicit notice to the contrary.

22. See footnote 3: 'Evolutionary theory moved the purpose which seemed to be displayed by the human genetic endowment from antecedent design to subsequent selection by contingencies of survival' (Skinner, 1976, p. 246).

23. Outside of theories of mind and theories of evolution, we know of only one other serious scientific theory that is constructed on a random-generator-and-filter model; namely, the classical theory of the market, according to which producers are generators of more-or-less random efficiencies and consumers choose among products in accordance with principles of a more-or-less rational decision theory. (Darwin was, of course, entirely alert to the similarity between market theory and his account of evolution.) Whether the analogy should be of comfort to adherents of either theory is very much open to dispute.

24. However, Skinner doesn't quite mean this, as he is also explicit that OT 'assumes the intact organism'. The latter formulation allows for the possibility of unconditioned (e.g. innate) features of the psychological profiles on which OT operates. Skinner assumes that if there are any such, they would themselves consist of S–R pairs. Pavlovian unconditioned reflexes would be paradigms.

25. As OT operates under the iterativity constraint, it will likewise apply to indefinitely many psychological profiles where the strength of S–R associations is arbitrarily larger than zero. These profiles will themselves be consequences of previous encounters with reinforcers.

26. Hume and Hull, by contrast, both thought that associative strength is subject to principles of 'contiguity' and 'similarity'. Although it now seems rather quaint, this issue was much argued in its time.

27. OT and ET must both reconcile gradualism with the thesis that the novel forms among which environmental variables select are generated at random. They do so in exactly the same way – by assuming that the randomness of generative processes typically consists of variation around an ancestral mean. So if G1 butterflies are reddish-brown, G2 butterflies may include some that are more brown than red and some that are more red than brown. Selection pressures may then choose among these, thus shifting the modal color of the G2 descendants in one direction or the other.

28. The question of just what 'expressing' amounts to is wide open, although it's clear that it's nothing like a one-to-one relationship: it is generally misleading to speak of 'the' gene 'for' such and such a phenotypic trait or 'the' phenotypic trait that is the expression of such and such a gene. For a nuanced approach, see Fisher (2006).

29. Parallel reasoning has been central to discussions well outside theories of learning and theories of evolution. Epistemologists have worried for centuries about how it could be that our contingent beliefs are largely true if, on the one hand, the world isn't mind dependent and, on the other hand, the mind is 'active' in the process of belief fixation. Kant's First Critique remains the most serious attempt to provide the required reconciliation, but few really think that it succeeded.

Chapter 2

1. Experimental data on, and computer simulations of, rather straightforward natural selection acting on random genetic mutations are being published in the most accredited biological

journals as of this writing (March 2009). See, for instance, Teo'tonio et al. (2009), just one instance among many.

2. In the words of an expert on the history of evolutionary theories: 'Darwin's theory of natural selection ... concentrates on the steady pressure of the environment when changes are very slow ... There can be no direction imposed on evolution by factors internal to the organisms, because the variation upon which selection acts is random in the sense that it is composed of many different and apparently purposeless modifications of structure. The environment determines which shall live and reproduce, and which shall die, thus defining the direction in which the population evolves ... In such a theory, the course of evolution is open-ended and unpredictable, because each population is subject to changes in its local environment or may encounter entirely new environments through migration' (Bowler, 2003, pp. 10–11). (We are grateful to Professor Thomas Lindell of the University of Arizona for this quote.)

3. The expression 'shifting balance' was introduced into evolutionary theories by Sewell Wright in 1932 (Wright, 1932). This is the component of 'invisible hand' that Darwin himself had probably imported from Adam Smith and the evolution of free markets. The expression appears in Smith's *The Wealth of Nations* and in *The Theory of Moral Sentiments* with slightly different characterizations.

4. A smart, simple software, called NetLogo, allows one to visualize, step by step, the evolution of model situations. A priori intuition cannot anticipate the outcomes.

5. Some caveats are needed here. No neo-Darwinian would propose that hill climbing would be just as likely to generate a new trait requiring one base change as to a generate a new trait requiring ten base changes. Traits do not emerge with equal probability from some uniform flat 'trait space'. Several probabilistic nuances and calculations were part of the canonical picture (Rice, 2004). But we think we can be excused from attending to these nuances here. The 'blind' character of

internal variation by mutation is indeed a prominent component of the neo-Darwinian picture (see especially Dawkins, 1986). Getting a lot (many highly complex life forms) from very little (a blind source of diversity) is supposed to be the glory of neo-Darwinism, the core (in Dennett's terms) of 'Darwin's dangerous idea' (Dennett, 1995) (the reader be cautioned: Dennett thinks it's a marvellous idea – perhaps the best anyone has ever had; see also Dennett, 2007). In short, internal variation by mutation is supposed to be strictly random with respect to the selectional environment; there is no 'look ahead' in deciding which mutations to produce. That much we are ready to grant, although epigenetics (see below) sometimes seems to suggest a rather different story.

6. An important technical problem worth mentioning here is that these representations assimilate the combination of gene frequencies to the frequency of their combination, and the rate of reproduction by a genotype to the rate of reproduction of that genotype (for a detailed critique of these confusions in formal models of Darwinian fitness, see Ariew and Lewontin, 2004).

7. For an application of fitness landscapes to (heaven help us) industrial supply networks, see Li et al. (2008).

8. For a recent balanced evaluation of Haldane's defence, see Ewens (2008).

9. In a remarkable early (1985) review of developmental constraints in evolution, we read: '[In adaptive evolution] it must often be the case that small changes in genotype often cause small changes in phenotype and that genetic changes altering one trait do not always result in excessively maladaptive changes in others. If complex adaptations involving many genes are to evolve by natural selection, it must be possible to change single traits without disturbing the others in a way that reduces overall fitness ... *However, there is more to evolution than an increase in adaptation, and various types of developmental constraints, linking diverse traits, do exist*' (Maynard Smith et

al., 1985, p. 266, emphasis added). We will return to this paper later on.

10. The origin of this story is detailed in the Wikipedia article 'Bert & I' (see http://en.wikipedia.org/wiki/Bert_and_I [accessed September 2009]).

11. We have been reminded recently (October 2008) of the unabated importance of DNA sequences in an editorial in Science, aptly entitled 'It's the sequence, stupid!' (Coller and Kruglyak, 2008). In a nutshell, the transplantation of an entire human chromosome (chromosome 21) into a mouse cell that also contains the whole complement of the mouse genome shows that it is the regulatory DNA sequence, rather than any other species-specific factor, that constitutes the single most important determinant of gene expression (Wilson et al., 2008).

12. This term was coined in the mid-1990s from the words 'evolution' and 'development', rightly seen as two faces of the same coin. It became almost at once successful. It's presently common currency both in biology proper and in scientific popularizations.

13. Excellent accessible expositions have been available now for some years: West-Eberhard, 2003; Kirschner and Gerhart, 2005; Carroll, 2005, 2006. An earlier important recognition came with the 1995 Nobel Prize for Physiology or Medicine to Lewis, Nüsslein-Volhard and Wieschaus (see Lewis et al., 1997).

14. There are at least 282 genes in humans that are somehow associated with mental retardation, of which 16% have orthologues (genes precisely corresponding, by position, sequence and function) in the fruit fly (Restifo, 2005). Detailed studies of the neural causes of specific forms of mental retardation in children (Noonan syndrome) are carried out by 'knocking out' corresponding genes in the mouse, and by studying the developmental role of analogous genes in the fruit fly (Gauthier et al., 2007).

15. As early as 1992, a transferred human master gene (HOX4B) was shown to produce a head-specific expression in *Drosophila* (fruit fly) embryos (Malicki et al., 1992).

16. A recent daring hypothesis, of a universal genome, has been advanced by Michael Sherman of Boston University. He says: 'I propose [the hypothesis of a] Universal Genome that encodes all major developmental programs essential for various phyla of Metazoa emerged in a unicellular or a primitive multicellular organism shortly before the Cambrian period ... The Metazoan phyla, all having similar genomes, are nonetheless so distinct because they utilize specific combinations of developmental programs. This model has two major predictions, first that a significant fraction of genetic information in lower taxons must be functionally useless but becomes useful in higher taxons, and second that one should be able to turn on in lower taxons some of the complex latent developmental programs, e.g., a program of eye development or antibody synthesis in sea urchin' (Sherman, 2007).

17. For the opposite polarity in the origins of the nervous systems in insects and vertebrates see Sprecher and Reichert (2003).

18. Much more in Chapter 8 against this ill-conceived notion that there are evolutionary 'problems' that species are supposed to have to 'solve'. (Germane considerations are to be found in Pigliucci 2009a, page 223.)

19. The first such laboratory reproduction of a plausible evolutionary event was Ronshaugen et al. (2002).

20. We want to mention here, but not develop, the close parallel between the notion that a variety of life forms have been generated by differences in regulatory processes acting on the same (or very similar) genes and the idea that the manifest differences in the languages of the world may have been generated by changes in a small set of 'parameters' acting on the same (or very similar) basic linguistic principles. (For a recent treatment see Chomsky, 2009; for a classic treatment, see Chomsky, 1981; for one of the earliest detailed cases, see

Rizzi, 1989). Chomsky's initial suggestion (in lectures at MIT in 1978 and 1979) was explicitly motivated by ideas due to the French geneticist (and Nobelist) François Jacob about how slight rearrangements of timing and organization of regulatory circuits might yield the difference between an elephant and a fly (see also Chomsky, 2009).

21. A uniquely human developmental factor (a gene enhancer called HACNS1) accounting for the particular formation of human limbs has been recently identified by comparison with the corresponding factor in transgenic mice, rhesus monkeys and chimpanzees (Prabhakar et al., 2008).

22. For recent comprehensive reviews, see Mattick, 2005; Amaral and Mattick, 2008; Mattick and Mehler, 2008; Stefani and Slack, 2008.

23. The messenger RNA undergoes several modifications and processing (notably splicing, see below) and is then 'translated' in the ribosomes into a sequence of amino acids, called a polypeptide. Finally the folding of this sequence into a specific three-dimensional configuration is the protein that is said to be 'coded' by the gene. The complex relation between the linear sequence of the amino acids and the three-dimensional protein configuration has been actively studied for decades but remains to this day elusive (for recent advances on the 'knotting' of the polypeptide chain, see Mallam et al., 2008).

24. It has been suggested recently, on solid evidence and sound theory, that the internal structure of living cells (bacteria excluded) is relevantly similar to the structure of glass (Trepat et al., 2007) (we are grateful to Professor Fernando Martinez of the University of Arizona for bringing this interesting idea to our attention). Further suggestions of a 'glass-like' structure of the fitness landscape in a mathematical model of speciation in general have been made by Heo et al. (2009).

25. See the next chapter on the pioneering experiments by Waddington.

26. A recent analysis of the buffering of mutations by HSP90 in the plant *Arabidopsis thaliana* has led the authors to state: 'This result strengthens the plausibility of previous suggestions that HSP90 might play an important role in enhancing the rate of evolutionary change. Moreover, we propose that far more genetic variation may be available to alter phenotype than quantitative genetic studies generally suggest' (Sangster et al., 2008, p. 2973). To the simple question of how many traits can be affected, their answer is 'nearly every one'. 'We demonstrate that HSP90-responsive natural genetic variation can be observed in *A. thaliana* at such a frequency that nearly every trait could be expected to be affected.'

27. Quite intuitively, supposing that a gene contains ten exons, in one case (typically, in one kind of tissue) all of them are transcribed and expressed in that order. In another case (in another kind of tissue) we can have, say, the sequence 1,2,5,6,7,8,9,10, in yet another 1,2,3,4,7,8,9,10 and so on. A mutation affecting one exon will affect all the proteins in which it is expressed, hence affecting many tissues at once.

28. A felicitous formula suggested by Schmucker and Chen (2009) when reviewing data on alternative splicing of a number of genes connected with the immune system, neural wiring and the adhesion between cells in various species is: 'complex genes in simple animals, complex animals yet simple genes'.

29. Alternative splicing also seems to be linked to a process (technically called 'polyadenylation') whereby a repeat of chemical groups (the adenines) are added to the ends of mRNA to help stabilize it. Both processes (alternative splicing and stabilization) appear to be regulated by a common mechanism.

30. BGC is not the only mechanism in Dover's original picture. There is also unbiased gene conversion, a process that has basically the same effects as BGC, but over much longer evolutionary times.

31. Notably Gerald Edelman, in several popularization books (e.g. Edelman, 1987) and technical articles, has promoted his

'theory of neuronal group selection' (TNGS). Similar 'internal' Darwinian mechanisms for the development of the brain have been proposed by French neurobiologist Jean-Pierre Changeux (Dehaene et al., 1987).

Chapter 3

1. An excellent short summary of gene regulation in higher organisms (eukaryotes) is to be found in http://users.rcn.com/jkimball.ma.ultranet/BiologyPages/P/Promoter.html [accessed September 2009].

2. Let's be reminded of the biological classifications, top to bottom, with some simple canonical examples for each group. Historically, it goes all the way back to Carl Linnaeus (1707–1778):
Domain (eubacteria, archaea, eukaryotes)
Kingdom (protista, chromista, fungi, plants, animals, metazoa)
Phylum (chordates, arthropods, molluscs, annelids, echinoderms)
Sub-phylum (urochordates, cephalochordates, vertebrates, crustaceans)
Class (mammals)
Order (primates)
Family (lemuridae)
Genus (lemur)
Species (ring-tailed lemur)

3. Interestingly, a debate ensued in *Science* pitting Davidson and Erwin against a defender of more traditional Darwinism, Jerry A. Coyne, who questioned these assumptions and claimed to be able to show that natural selection adequately explains the origin of new phyla. Coyne characteristically (and perplexingly) states that 'it is unlikely that macromutations played an important role in major evolutionary change' (Coyne, 2006). See the reply by Erwin and Davidson (2006). See also Coyne (2009).

4. Of course, nobody ever claimed that variations arise completely at random. As we said earlier, even in the canonical neo-Darwinian picture, a variation that is the result of, say, ten mutations, is less likely to arise than one that is the result of one mutation only. Such differential probabilistic patterns have been assumed as a standard. However, the 'blind' character of mutations and some kind of random 'engine' at the roots of genetic variability are at the very core of the picture.

5. It doesn't, however, tell us that the mechanism by which it evolved was natural selection. Darwinism implies environmentalism, but not conversely. This is important because, as we'll see in later chapters, there remains something deeply wrong with selectionism even if environmentalism is assumed.

6. This is strictly connected with the notion of evolutionary spandrels, to which we will amply return in Chapter 6.

7. Interesting considerations are developed by these authors, based on experimental data and accurate mathematical models. 'The most consistent result in more than two decades of experimental evolution is that the fitness of populations adapting to a constant environment does not increase indefinitely, but reaches a plateau. Using experimental evolution with bacteriophage, we show here that the converse is also true. In populations small enough such that drift overwhelms selection and causes fitness to decrease, fitness declines down to a plateau. Both of these phenomena must be due either to changes in the ratio of beneficial to deleterious mutations, the size of mutational effects, or both. The most significant change in mutational effects is a drastic increase in the rate of beneficial mutation as fitness decreases. In contrast, the size of mutational effects changes little even as organismal fitness changes over several orders of magnitude (a factor 300 in this case)' (Silander et al., 2007). The overall size of the population is shown to be crucial in determining the possibility of fixation or disappearance both of advantageous and of disadvantageous mutations.

8. See Schlosser (2004).

9. A cogent defence of the idea that dynamic developmental modules may have been the key to the explosion of new life forms in the Cambrian is to be found in Newman and Bhat (2008).

10. We see that the notion of developmental and evolutionary modules isn't really so simple to define, but it's easy to get it intuitively. Let's say that components belong to a module when it is easier for them to communicate with one another than with anything else. Think of a jury that has a spokesperson: the members of the jury can talk to one another, but only the spokesperson can talk to the judge.

11. In the fruit fly, an antenna may be made to grow where normally a leg develops, or the rudiments of an eye can be made to develop almost anywhere, by activating the *Pax6* gene complex.

12. The BMP4 protein (a ubiquitous bone morphogenetic protein) and the factor CaMKII (calmodulin kinase II).

13. As usual, we aren't saying that we've got right what environmentalists get wrong. Our point is just that, contrary to frequently heard claims that neo-Darwinism is the only game in town, there are, in fact, prima facie plausible alternatives. Nobody knows exactly how evolution works, so internalists sometimes tell 'just so' stories too.

14. 'Although evolutionary change, biogeographic provinciality, and paleo-environments might have played a role in Ediacara taxonomic evolution, they do not seem to have controlled the overall range of the realized morpho-space, which appears to be invariant to notable taxonomic differences. Thus, changes in taxonomic diversity that occurred through time while morpho-space range remained relatively constant should affect the internal structure of morphospace' (Shen et al., 2008).

15. A recent vindication of discontinuous evolutionary processes (an explicitly and intentionally dissonant note in the vast concert of celebrations of the Darwinian bicentenary) is found in Theissen (2009). Let's be reminded of the oft-quoted

and rather sad (in hindsight) passage in Darwin that we have transcribed at the very opening of this book: 'If it could be demonstrated that any complex organ existed which could not possibly have been formed by numerous, successive, slight modifications, my theory would absolutely break down' (Darwin, 1859, p. 194).

16. We are, once more, indebted to Richard Lewontin for having rightly insisted, over many years, on the crucial importance of the transitivity assumption in adaptationist explanations.

17. 'Epistasis means that the phenotypic consequences of a mutation depend on the genetic background (genetic sequence) in which it occurs ... These distinctions are crucial in the context of selection. Mutations exhibiting magnitude epistasis or no epistasis are always favoured (or disfavoured), regardless of the genetic background in which they appear. In contrast, mutations exhibiting sign epistasis may be rejected by natural selection, even if they are eventually required to increase fitness. In other words, some paths to the optimum contain fitness decreases, while other paths are monotonically increasing. When all paths between two sequences contain fitness decreases, there are two or more distinct peaks. The presence of multiple peaks indicates reciprocal sign epistasis, and may cause severe frustration of evolution. Indeed, reciprocal sign epistasis is a necessary condition for multiple peaks, although it does not guarantee it: ... two may be connected by a fitness-increasing path involving mutations in a third site' (Poelwijk et al., 2007, p. 383).

18. Richard Lewontin aptly insists on the greater value of this different metaphor.

19. A genetic analysis of the many-tentacled polyp Hydra (also rich in mythological overtones), carried out with sixteen different types of Hydra from fifteen widely separated and ecologically distinct localities in India, has revealed that local adaptation is simply the result of switching specific genes, depending on the ecological requirements (Rastogy and Pandey, 1992; Hoenigsberg, 2002).

20. The special implications of such systems for the action of natural selection had long been recognized by Sewall Wright (Wright, 1931).

21. In their reply to Coyne, Erwin and Davidson emphatically declare: 'Nowhere in our paper did we reject natural selection, because we support it'.

22. There is at present only one rather well-supported adaptationist reconstruction of human phenotypes: protection from malaria conferred by the heterozygotes for haemoglobin S. A much less plausible case is the one of vitamin D, and fair skin related to poor exposure to sunlight. Contrary to many claims made by evolutionary psychologists, no clear case exists at all in the domain of human behaviours and cognition (for germane considerations, see Lewontin, 1998).

Chapter 4

1. These variants are aptly called ultrabithorax, and the master gene controlling this developmental pathway is called *Ubx*. Interestingly, as a further instance of the interchangeability of reactions to external shocks and to genetic changes, the phenotypes resulting from well-characterized (homeotic) mild mutations in the *Ubx* gene closely resemble the phenotypes resulting from early exposure to the ether fumes.

2. We wish to express our debt to Richard Lewontin for suggestions on this whole section.

3. Later work by Rendel (Rendel, 1968, 1969) – especially focused on the 'scute' mutant in the fruit fly – has shown that there are genes that buffer development, in the sense that small internal variations and small variations in an ordinary environment have no effect downstream in development. The shock, being a major variation in the environment, uncovers the effects of otherwise hidden variation, driving a few individuals outside the zone of buffering. Some buffering genes, somewhere in the genome of those individuals, are not doing such a good job. With

repeated selective breeding, the frequency of those gene variants increases. More and more individuals are completely outside the buffering zone and they can produce abnormal offspring without need of any shock. Rendel and Lewontin point out that this is the explanation of what Waddington had aptly called 'canalization' and 'genetic assimilation'.

4. Claims can be made that we have not yet discovered the adaptive role of these phenotypes. Adaptationists are inclined to make such claims. But we wish to suggest that it's likely that these phenotypes have no adaptive role at all. The complex dynamics of repeated selection and cross-breeding in the laboratory, over many generations, unmasks pre-existing silent mutations, and the different developmental pathways induced by these give rise to new phenotypes. No adaptive process is needed. Analogous phenomena have no doubt happened in evolution.

5. We will return to this issue in Chapter 8.

6. For a full mathematical treatment see Rice (2004).

7. We are indebted to Richard Lewontin for drawing our attention to this case and for comments and suggestions on an earlier draft of this section.

8. Only two out of the four initial strains could ferment (digest) sugar 2. These were taken and a further selection for sugar 3 was tried. One strain was such that this could never be obtained. In essence, there were several dead ends after the first phase, and one dead end after the second phase.

9. Another remarkable case of the role of contingency in evolution has been published recently (Blount et al., 2008). After monitoring 31,500 generations in 12 identical populations of E. coli, a variant capable of utilizing citrate finally appeared. These authors conclude that it is probably an ordinary mutation, but one whose physical occurrence or phenotypic expression is contingent on prior mutations in that population. They say: '... the evolution of this phenotype was contingent on the particular history of that population. More generally, we suggest that historical contingency is especially important

when it facilitates the evolution of key innovations that are not easily evolved by gradual, cumulative selection' (p. 7899). The importance of contingency, now increasingly stressed in biology proper, supports our suggestion that evolutionary explanations based on adaptation and natural selection are not in the same league as proper scientific laws, but rather in that of historical explanations (see Chapter 9) (D'Arcy Wentworth Thompson had insightfully suggested exactly this, back in 1917 in his monumental essay 'On growth and form' [Thompson, 1917].)

10. The size of one of these chemical groups that can become stuck to the histones, or to DNA, is to the size of the histones themselves, or of the DNA composing an entire gene, as the size of a pebble picked up by a groove in a tyre is to the size of a whole bus. But small does not mean unimportant.

11. A detailed genetic and epigenetic analysis of the DNA sequences in various cells of identical (monozygotic) and fraternal (dizygotic) twins, with special attention to the differences in the DNA methylation sites, has been completed (February 2009) in concert between laboratories in Canada, the USA, Taiwan and Sweden. Cutting a long and fascinating story quite short, the 16 authors of the report conclude 'that molecular mechanisms of heritability may not be limited to DNA sequence differences' (Kaminsky *et al.*, 2009).

12. If you are of Swedish origins, you had best hope that your grandfathers didn't have too much to eat, but that your grandmothers did.

13. Jean Baptiste de Lamarck (1744–1829) maintained that evolution, and speciation in particular, takes place via inheritance by the offspring of individual modifications caused by the environment in their parents. Although Darwin was a kind of self-confessed Lamarckian, Darwinism sounded the death knell for any such idea, and it is still the case that the simple mention of Lamarckism leads the majority of biologists to draw their revolvers. Whence the present perplexity caused by some aspects of epigenetics. Let's state clearly that, at present,

only quite tentative links have been established between the inheritance of epigenetic traits and speciation (Randy Jirtle, personal communication, February 2009). We are bringing epigenetics into our picture because it's an extremely vital new field of inquiry, because it offers a remarkable instance of the multiplicity and subtlety of different 'environments' and to stress the many nuances that one has to take into consideration when connecting genes to phenotypes (for early reviews, see Hoenigsberg [2002] and Pembrey [2002]).

14. A recent (October 2008) analysis of one human gene, *SEPN1*, which is known to be involved in a type of muscular dystrophy, along with comparative data from chimpanzee and macaque tissues, suggested that the presence of a muscle-specific Alu-derived exon resulted from a human-specific change that occurred after humans and chimpanzees diverged evolutionarily (Lin *et al.*, 2008).

15. In a recent report (October 2008) a team of Stanford geneticists has explored the population dynamics and evolution of transposable elements in the fruit fly (*Drosophila*), with special attention to their possible adaptive role. Theirs is the first comprehensive genome-wide screen for recent adaptive TE insertions in *Drosophila*. Using several independent criteria, they identified a set of 13 'adaptive' TEs. It is estimated that 25–50 TEs have played adaptive roles since the migration of *Drosophila* out of Africa. These TEs are judged by these authors to have contributed significantly to local adaptation in this species (González *et al.*, 2008).

16. In particular: the development of the brainstem trigeminal nuclei (the so-called barrelettes), of the barreloids in the thalamus and of the barrels in the neocortex.

Chapter 5

1. 'From above' doesn't, of course, mean 'from God'; it means multi-molecular and multicellular factors and abstract formal principles. Nothing else.

2. Technical note: in the Fibonacci series, each term is equal to the sum of the two preceding ones (1, 1, 2, 3, 5, 8, 13, 21, and so on). Connecting the outer vertices of an ordered pattern of juxtaposed squares that have areas given by the Fibonacci series with a continuous curve, we obtain the Fibonacci spiral. The ratio between two successive terms of the Fibonacci series tends to the golden mean as a limit (approximately 1.61803399). Fibonacci spirals are usually formed when the elements of a pattern optimize their disposition with respect to two opposing forces. The presence of Fibonacci patterns is ubiquitous in plants (phyllotaxis) (Maynard Smith et al., 1985), and two French statistical physicists, Stéphane Douady and Yves Couder, have shown how these arise in nature, in a laboratory experiment (with magnetically charged droplets) and in mathematical simulations, from self-organization in an iterative process. These patterns, realizing optimal packaging solutions, depend only on initial conditions and one parameter that determines the successive appearance of new elements. The ordering is explained by the system's tendency to avoid rational (periodic) organization, thus leading to a convergence towards an angle dictated by the golden mean. For beautiful figures and a formal treatment, see Douady and Couder (1992). For a movie clip showing the formation of Fibonacci spirals by the droplet in real time, see http://maven.smith.edu/~phyllo/Assets/Movies/DouadyCouderExp5.9MB.mov [accessed September 2009].

3. Logarithmic spirals are commonly observed in molluscs, brachiopods and some foraminifera, as remarked already by D'Arcy Wentworth Thompson (Thompson, 1917, 1992) and later analysed mathematically and empirically by David M. Raup, Steven Jay Gould and A. Michelson (for an analysis and a

rich bibliography, see the already cited review by Maynard Smith et al. [1985]; see also Raup [1966, 1967]).

4. Zexian Cao and colleagues at the Chinese Academy of Sciences recently used stress engineering to create differently shaped microstructures just 12 μm across with a silver core and a SiO_2 shell. They discovered that if the shells were encouraged into spherical shapes during cooling, 'golden' triangular stress patterns formed on the shells. On the other hand, if they were encouraged into conical shapes, spiral stress patterns were formed. These spiral patterns were Fibonacci spirals. Their comment is that biologists have long suspected that the branching of trees and other occurrences of the Fibonacci sequence in nature are simply a reaction to minimize stress; they say that their experiment 'using pure inorganic materials may provide proof to this principle' (Cartwright, 2007; see also Li et al., 2007).

5. For (what some may consider rather daring) applications of the Fibonacci numbers to linguistic structures at various levels, see Medeiros, 2008; Piattelli-Palmarini and Uriagereka, 2008; Soschen, 2008.

6. It is of some historical interest that the great German poet and naturalist Johann Wolfgang von Goethe, inspired by Plato's theory of eternal and changeless forms, and by Spinoza's doctrine of an infinite combination of 'modes', had the idea of Urpflanze, the archetypal forms after which all other plants are patterned. However, the modern scene actually starts with D'Arcy Wentworth Thompson.

7. He suggested, as we also do in Chapter 7, that evolutionary explanations are historical and narrative in character, employing the same intentional and teleological vocabulary we use in presenting human history, and hence, while perhaps on occasion of heuristic value, they are not part of biology as a natural science (see also Leiber, 2001).

8. Some of Turing's statements in that paper sound rather preposterous today: '... it is only by courtesy that genes can

be regarded as separate molecules. It would be more accurate (at any rate at mitosis) to regard them as radicals of the giant molecules known as chromosomes ... The function of genes is presumed to be purely catalytic. They catalyze the production of other morphogens, which in turn may only be catalysts.'

9. An interesting anecdote: in 1951 Belousov (Director of the Institute of Biophysics in Moscow) submitted a paper to a scientific journal reporting to have discovered an oscillating chemical reaction. It was roundly rejected with a critical note from the editor that it was clearly impossible. The editor's confidence in its impossibility was such that even though the paper was accompanied by the relatively simple procedure for performing the reaction, he could not be troubled. If citric acid, acidified bromate and a ceric salt were mixed together, the resulting solution oscillated periodically between yellow and clear. He had discovered a chemical oscillator. (See the website of Rubin R. Aliev, Institute of Theoretical & Experimental Biophysics, Puschino, Moscow Region, Russia, and movie clips of such reactions at http://online.redwoods.cc.ca.us/instruct/ darnold/DEProj/Sp98/Gabe/intro.htm [accessed September 2009].)

10. Catastrophes in systems with only one state variable: the fold (one control parameter); the cusp (two control parameters); the swallowtail (three control parameters); the butterfly (four control parameters). Catastrophes in systems with two state variables: the hyperbolic umbilic (three control parameters); the elliptic umbilic (three control parameters); the parabolic umbilic (four control parameters). Thom proved that no classification can be made at all for systems with more than four control parameters.

11. An exception is the attention to, and endorsement of, Waddington's work expressed in the 1985 review by Maynard Smith, Burian, Kauffman, Wolpert and colleagues (Maynard Smith et al., 2005). The work of Rendel and his Australian school, as mentioned in the previous chapter, offered an

explanation of the phenomena discovered by Waddington, and his model of genetically modulated developmental buffering is a bridge to the modern field of epigenetics.

12. Mostly, it has to be said, by stressing the importance of the laws of form for evolution and development rather than offering workable concrete models.

13. Maynard Smith and Savage (1956) stressed how the law of the lever requires that any uncompensated changes in the speed with which a limb can be moved will reduce the force that it can exert.

14. Lewontin has recently expressed perplexity towards the terminology of 'laws of form', being doubtful that there are any genuine 'laws' in biology (personal communication, October 2008).

15. Eva Jablonka and Marion Lamb, in a 2005 book that explains very clearly most of the recent developments in evolutionary biology and rightly pleads for a radical reconsideration of evolutionary theory, completely ignore the issue about the laws of form (see Piattelli-Palmarini, 2008). Curiously, they stress the need for a 'fourth dimension' in evolution, ignoring that West, Brown and Enquist in 1999 had introduced this very expression for a totally different aspect of evolution (the fractal law) (West *et al.*, 1997, 1999, 2002).

16. The natural general equation is of the form $Y = Y_o(M)^b$, where b is the scaling exponent, M the body mass and Y_o a normalization constant. It turns out that b is a simple multiple of ¼. For instance:
    - diameter of tree trunks and aortas, $b = ⅜$ (therefore, for their cross-sectional area, $b = ¾$)
    - rates of cellular metabolism and heartbeat, $b = -¼$
    - blood circulation time and lifespan, $b = ¼$
    - whole organism metabolic rate, $b = ¾$.

17. The field matured in the 1970s for microcircuit design, typically to minimize the total length of wire needed to make a given set of connections among components.

18. The nearly optimal character of the genetic code is another instance. Among thousands of possible alternatives, the genetic code as we know it is optimal for minimizing the effect of frame-shift mutations and minimizing the energy wasted in synthesizing the start of anomalous protein sequences. In the words of the authors: 'the universal genetic code can efficiently carry arbitrary parallel codes much better than the vast majority of other possible genetic codes' (Itzkovitz and Alon, 2007).

19. In the case of brain connectivity optimization *à la* Cherniak and colleagues we have a rather precise calculation of the hypothetical search space. The distribution of wirecosts (total wirelength) of all possible layouts of ganglia of the 'simple' nematode *Caenorhabditis elegans* (a 10,000-bin histogram) represents almost 40 million possibilities (39,916,800 alternative orderings). The numbers for the possible layouts of the nervous systems of more complex and more recent species spiral upwards steeply. The possibility of a blind search, followed by natural selection, in such gigantic spaces is extremely implausible.

20. As stated elsewhere (in Part two), we use here for mere expediency this (alas) standard notion of 'problems' posed to the evolution of organisms, and of 'solutions' to these problems, with scare quotes.

21. We are indebted to Professor Marc Hauser of Harvard University for presenting these data by Bejan and Marden at a symposium in 2005, before they were published.

22. They say: 'The long evolution of vascular plants has resulted in a tremendous variety of natural networks responsible for the evaporatively driven transport of water. Nevertheless [until now], little [wa]s known about the physical principles that constrain vascular architecture' (Noblin *et al.*, p. 9140).

23. The specialized literature on optimal foraging is huge, spanning the individual and collective behaviours of a variety species of fish, ants, bees, birds, deer, monkeys and apes.

24. Both Anna Dornhaus and Richard Lewontin pointed this out to us in personal exchanges.

25. Let's stress here and now that the quite popular metaphor of 'problems' that require 'solutions' in the domain of evolution is badly misguided (see Part two for a detailed discussion, see also Lewontin [2000] and Pigliucci [2009a]). We have used it previously and we use it again here only for the sake of simplicity, but our present considerations do not depend on taking this metaphor seriously. Its use can only worsen the case of adaptationist neo-Darwinism.

26. The term 'satisficing', initially coined in the domain of decision-making by the economist and psychologist Herbert Simon (1916–2001), winner of the Nobel Prize for Economics in 1978, characterizes a strategy which, somewhat more modestly and more rapidly, attempts to meet criteria for adequacy, rather than to identify an optimal solution. Under this or similar labels, the concept has been widely adopted by evolutionary biologists such as John Maynard Smith, by neo-Darwinian cognitive scientists such as Daniel Dennett and Gerd Gigerenzer (fast and frugal heuristics) and by researchers in artificial intelligence and computer science.

27. Evolutionary biology has traditionally been concerned with explaining why there are the life forms that there are. By contrast, the present issue is why there aren't the life forms that there aren't (see the previous chapter). We will return to such issues in Chapter 7.

28. For example, in the helical morphology of colonies of moss animals called bryozoans, a very common fossil the world over, actual forms cluster into a handful of shapes, showing a space that is otherwise massively empty (McKinney and McGhee, 2003).

Chapter 6

1. Terminological note: in what follows, we'll sometimes say things like 'white bears were selected-for their colour', and sometimes we'll say things like 'whiteness was selected-for in polar bears'. These are meant to be synonyms; merely stylistic variants. However, both contrast with 'white polar bears were selected'; the first two are interchangeable with one another, but not with the third.

2. The issue of what fitness is is notoriously controversial (for an interesting criticism of the standard concept, see Ariew and Lewontin, 2004; for a mathematical treatment, see Rice, 2004). But it will do for our purposes to assume, as adaptationists generally do these days, that whatever fitness consists of, it is proportional to a creature's likelihood of reproducing. Assuming this eases the exposition, but our arguments won't depend on adopting one definition of fitness rather than another.

3. The present section may be of special relevance to our main purposes, since we've often been told that the whole of Biology rests on the notion of teleological function, which is in turn kept aloft by the Darwinian account of evolution; to a first approximation (but see Gould and Lewontin, 1979), the function of a biological organ is whatever function it was selected-for performing. We are, however, deeply suspicious of this line of thought. For one thing, the indispensability to Biology of the notion of biological function, although widely asserted, has not, to our knowledge, been widely argued for. It's a topic somebody ought to write a book about. Second, the Darwinian account of biological function has the striking disadvantage of being diachronic; what function one's heart has now depends on what function it was selected-for millions of years ago; with, it appears, the peculiar consequence that if Darwin proves not to be right about the contingency of evolution on selection, it would follow that one's heart has no function. Still, it's surely true that if Darwin's story about the

role of selection in the evolutionary fixation of phenotypic traits is untenable, then currently standard accounts of teleology, and its indispensability to Biological theories, will be due for significant reconsideration. That strikes us as not at all a bad thing.

4. More about Mother Nature in Chapter 7. Please assume, for now, that she's just a way of talking.

5. Psychologists who were deeply immersed in the positivist epistemology that generally went along with learning theory were sometimes wont to claim that there is no fact of the matter about what the organism learns when it forms a conditioned S–R association (or that there is no fact of the matter until somebody actually runs an experiment that splits the stimulus; a psychological analogue, perhaps, to the equivocal situation of Schrödinger's cat). Here, for example, is Howard Kendler, writing in 1952: 'There would be no confusion about the meaning of such terms if it were always remembered that these intervening variables serve as *economical devices* [sic]. They are '*shorthand' descriptions* [sic] and nothing more of the influence on behavior of several independent variables … the construct of learning, whether it be conceived in terms of modifications in cognitive maps or S–R connections, does not refer to any object thing or entity as suggested by those who are concerned with the question of what is learned' (Kendler, 1952, p. 271). It's striking that, having swallowed all that dubious methodology, we still don't know how Kendler wants to describe 'the several independent variables' in play in a conditioning experiment. Was the stimulus to which the pigeon was conditioned a triangle or a yellow triangle or something yellow? Kendler's paper makes fascinating reading, if only as a reminder of how much of a mess a psychologist can get into by believing what some philosopher tells him. 'Present-day philosophy of science has been concerned with establishing criteria for distinguishing between meaningful and meaningless questions' (ibid., p. 269). Any time now; but don't call us, we'll call you.

6. The story about what variables can affect what is learned in a conditioning experiment is extraordinarily complex; indeed, it is still a crux for experimentalists. For insightful analyses, see Gallistel, 2000, 2002.

7. We first heard about this neat little thought experiment from the late Professor Charles Osgood about 200 years ago.

8. And not just learning theory. See the discussion of appeals to the 'sleeping dog' heuristic in artificial intelligence in Fodor (2008).

9. Cognitivist theories of learning, unlike Skinnerian ones, assume that the question of 'what is learned' in S–R conditioning is equivalent to the question how the learner mentally represents the stimulus and the response. (Is the stimulus represented as a triangle or as something yellow? Is the response represented as a turn to the right? As a turn to the east? And so on.) Note once again the similarity to the arch and spandrel problem. The spandrel is the free-rider because it was the arch that the architect selected-for; what he had in mind was that the arches should hold up the dome. The role of appeals to mental representation in resolving selection-for problems will presently become one of our main themes.

10. Not, however, to be confused with Saul Kripke's kind of causal theory of (referential) content, which is explicitly non-reductive. The account he sketches of how names refer presupposes intentional notions like the 'baptismal intentions' of speakers/hearers. We wouldn't suppose that Kripke feels much enthusiasm for a naturalist programme.

11. That is, the assumption that, either in this part of the woods or in general, all and only the ABNs are flies.

12. Likewise, in a world in which all and only flies are ABNs, which of the two a frog snaps at (which is, as one says, the 'intentional object' of the frog's snaps) does not affect the number of flies that the frog gets to eat. If all and only flies are ABNs, a disposition to snap at flies will add to your overall fitness exactly as much as a disposition to snap at ABNs; not one fly more, nor one fly less.

13. It can't, of course, be correspondence to how things are in the real (i.e. the actual) world; by definition, *no* counterfactual corresponds to how things are in the real world.

14. Notice that if you don't have a notion of 'a trait that's selected for' then (a fortiori) you don't have a notion of trait selection, so you can't state the fundamental Darwinian thesis: that creatures have the traits they do because those traits are selected-for their connection with fitness. This undermines the thought that you might fix up standard selection theory by just not worrying about selection-for traits. It may be that Gould and Lewontin had some such ameliorist modification of strict Darwinism in mind. If so, they much underestimated the trouble that spandrels raise for adaptationism.

15. For present purposes, the 'causal role of a property' is the set of things that causes it to be instantiated together with the set of things that its being instantiated causes.

16. In the philosophical jargon: SFPs turn up when a causal theory (or explanation, etc.) employs 'intensional' contexts. A context is intensional if the substitution of coextensive terms is not truth-preserving in that context. Suppose all the Fs are Gs and vice versa. Then context C is intentional if substituting a term that refers to one for a term that refers to the other need not preserve truth. The classic examples include ascriptions of 'propositional attitudes'. It can be true that 'John admires Cicero' and false that 'John admires Tully' even though 'Cicero' and 'Tully' are coextensive names (they refer to the same Roman). Intentional states are intentional and ???

17. Not, anyhow, if it is legitimate to identify a creature with a bundle of traits, each of which has a characteristic effect on fitness in the creature's ecology. Many evolutionary theorists have cautioned that the effect a phenotypic trait has on a creature's fitness typically depends a lot on what other phenotypic traits it interacts with, so that talk of 'the selection of traits for their effect on fitness' implies an illicit abstraction.

We wouldn't be very surprised if such objections turned out to be sound. We'll return to the issue in Chapter 8.

18. Darwin was particularly struck by this in his investigations in the Galápagos. There are markedly different populations of flora and fauna even on nearby islands. Presumably that's because each such population was far enough from the others to ensure their causal isolation.

19. That an actual effect can't have a counterfactual cause is a special case of the scholastic maxim that an effect can't have 'more reality' than its cause. St. Thomas of Aquinas thought that proves that there is a God; but it doesn't.

20. Ironies abound. We think it's likely that Darwin missed the significance of selection-for problems for the theory of natural selection because he was seduced by the putative analogy to artificial selection. In effect, his adaptationism was built on analogy to a case of intelligent design, viz. the intelligent design of phenotypes by breeders.

Chapter 7

1. There are lots of other ways of understanding the central role that considerations of counterfactual support play in the evaluation of empirical theories. Here's one: in typical cases, scientific predictions are generated by applying empirical generalizations (and/or empirical models) to specifications of initial conditions. These two factors are supposed to be independent in that the generalizations (or models) hold across a range of possible initial conditions, only some subset of which are actual. If this picture is right, then any theory that generates explanations or predictions of actual outcomes must willy-nilly generate explanations or predictions about counterfactual outcomes as well. Theories that seek to do the one can't opt out of doing the other.

2. Such examples aren't by any means merely hypothetical. Here's a famous example: domestic animals are bred for (inter alia)

their tolerance of interactions with the breeders; nobody wants a house cat that carries on like a panther. But it turns out, quite unexpectedly, that tolerance for interactions with breeders is linked to a whole lot of other phenotypic properties; so if you breed for domesticity, you're likely to get these other traits as free-riders. That's why there are respects in which domestic animals, of whatever species, tend to be more like one another than they are like their feral relatives. Dmitry K. Belyaev and collaborators have observed the appearance of dwarf and giant varieties, piebald coat colour, wavy or curly hair, floppy ears and other traits in domesticated varieties of many diverse species (sheep, poodles, donkeys, horses, pigs, goats, mice, guinea pigs and more) (Trut, 1999).

3. Compare a small but consequential slip that Sober makes (Sober, 1993, p. 18). '[An] obstacle that Darwin had to overcome [in using artificial selection as a model for natural selection] was consciousness [*sic*]. Artificial selection is the product of intelligent manipulation. Why think that organisms could be adapted to their environments without this sort of guidance?' But the relevant consideration isn't either that the process of artificial selection is intelligent or that it's conscious; it's that the process of artificial selection is intensional. Perhaps God is stupid or Granny is a zombie (that is, she has no conscious intentional states). The logic of the situation remains unchanged so long as her selections are performed with an end in view. We think the failure to keep these distinctions clear is an important reason why the kinds of problems about natural selection that concern this book have gone so generally unnoticed.

4. Or, rather, it favours them in 'nearby' counterfactual worlds. Presumably natural selection prefers fly-or-ABN-snappers to fly-snappers in worlds where ABNs are edible.

5. Designs do need designers, of course. Designs are such in virtue of their intentional histories, rather like Rembrandts. Accordingly, an accidental design is a sort of oxymoron.

Compare 'it just came out looking that way' with 'it just came out designed that way.' (By contrast, 'It just came out looking designed that way' is perfectly fine.)

It strikes us, by the way, that the case against 'intelligent design' (ID) explanations is even stronger than it is usually made out to be. It is sometimes said, rightly, that ID theories make no testable predictions. The Academy of Science 2008 booklet 'Science, Evolution and Creationism' states that 'Intelligent design is not a scientific concept because it cannot be empirically tested'. 'But so what?' comes the reply; 'quite likely string theory doesn't make any predictions that can be empirically tested either; but physicists seem to take it pretty seriously all the same.' There is, however, a reply to this reply: ID makes no predictions at all, testable or otherwise; all of its predictions are post hoc. The trouble is that there is no telling in advance what kind of world an intelligent designer might opt for. Maybe an intelligent designer might even opt for this world (but also, maybe not). Excepting only its logical consequences, nothing at all follows from ID. (Leibniz's argument that this is the best possible world is post hoc in exactly the same way, hence subject to the same objection.)

6. Even so, but that repetition dulls sensibility, one should find the proliferation of such theories very puzzling. Nobody thinks it would be a good idea to postulate a 'Granny Gravity' whose preference for objects that accelerate at thirty-two feet per second squared when they are unsupported in a vacuum explains why there are so very many objects that (would) do so. Nor are we urged to marvel at the ingenuity with which these objects have 'solved the problem' of falling in exactly that way in exactly such circumstances. What, one might well wonder, is supposed to be the salient difference between Granny Gravity and Mother Nature? (See also Chapter 3.)

7. For example, it belongs to (as one used to say) 'the logic' of intentions that they are sometimes thwarted. Well, could

Mother Nature intend to select for Fs but fail to do so? Do such failures depress her a lot?

8. 'But just how is it that intentional processes are sensitive to counterfactual outcomes if (merely) causal processes are not?' And why can't natural selection be sensitive to counterfactual outcomes in that way too, whatever way it is?' A very good question; to which, however, we don't know the answer. But we think it has something to do with, on one hand, the fact that intentional systems respond to states of affairs as (mentally) represented; and on the other hand, that there needn't actually be ABNs that aren't flies in order that there should be (mental) representations of ABNs that aren't flies. If that's right, then all we need is a theory of (mental) representation and everything will be clear. This sort of intuition has at least a respectable provenance. Thus Seager's (Strawson, 2006, p. 131) exposition of Leibniz: '… it is only via mental representation [*sic*] that an entire world can be wrapped up inside a single individual so that all [the] relations can be "read off" the intrinsic properties of that individual.' This strikes us as exactly right, and not just of thoughts about the actual world but of thoughts about counterfactuals as well. We don't, in fact, have a theory of representation (mental or otherwise). Fortunately, however, we don't need one for our present purposes, so long as we can assume that there are mental representations, and that they do have causal powers.

9. Speaking as fully signed-up atheists, we can't see much difference between claiming that God selects for fit phenotypic traits and claiming that Mother Nature does. So we do find it puzzling that many of our co-non-religionists insist on that distinction with such vehemence.

10. Admittedly, the tactic of resorting to scare quotes when push comes to shove (as in 'what natural selection "prefers"', 'what Mother Nature "designs"' or 'what the selfish genes "want"') can make it hard to tell just what is being claimed in some of the canonical adaptationist texts. Still, there are plenty of

apparently unequivocal passages. Thus Pinker (1997, p. 43): 'Was the human mind ultimately designed to create beauty? To discover truth? To love and to work? To harmonize with other human beings and with nature? The logic of natural selection gives the answer. The ultimate goal that the mind was designed to attain is maximizing the number of copies of the genes that created it. Natural selection cares only about the long-term fate of entities that replicate …' Fiddlesticks. The human mind wasn't created, and it wasn't designed, and there is nothing that natural selection cares about; natural selection just happens. This isn't Kansas, Toto.

11.   For another example: '[Darwin] argues by example, not analogy; the point of the opening of The Origin isn't that something similar happens with domesticated breeds and natural species; the point is that the very same thing happens, albeit unplanned and over a much longer period' (Gopnik, 2006, p. 56.). You might have thought that the caveat deserves some explication; how could a studied decision to select for one or other trait be 'the very same thing' as the unplanned culling of a population? If that's not just an analogy, what would be? Gopnik doesn't say.

12.   Which isn't to say that it's just an empirical issue. You'd think there might be a short refutation of the idea that there are laws of selection: namely that if there are (and if there is a random generator of phenotypes; see Chapters 1 and 2) then at the limit, creatures should be as similar as their environments; which they clearly aren't. The very same chunk of space–time can be inhabited by hundreds (not to say thousands; not to say tens of thousands) of species. There is, however, a reply to this: whether creatures should end up being similar depends not on the environment but on the ecology; and that creatures share a chunk of space–time doesn't imply that they share an ecology. But there is a reply to this reply: namely that the notion of an ecology (shared or otherwise) can't be specified unless such notions as selection pressures, competitions, etc., are already

assumed (see the discussion in Chapter 8). This poses a dilemma for the adaptationist; empirical implausibility on the one hand or circularity on the other. Take your choice.

13. Some biologists have claimed to descry 'trends' in evolution: that is, traits which will be selected-for in almost any ecological situation. But if there are such, they must be very coarse grained (in the literature, increased size, increased complexity and evolvability are typical candidates). And, as Mayr remarks (2001), 'almost all trends are not consistently linear, but change their direction sooner or later, sometimes repeatedly, and they may even totally reverse their direction' (p. 218).

14. For evidence that enhanced social signalling, rather than camouflage, seems to have been the driving force in the evolution of colour change in chameleons see Stuart-Fox and Moussalli (2008).

15. Laws govern relations between traits (properties). Accordingly, if there are laws of evolution, they must determine which traits win which competitions in which ecological situations. But, of course, traits are abstract; they compete only in the sense that creatures that have them do. It's creatures, not traits, that actually do the living and dying (compare Dennett, 2007). This ontological distinction between traits and the creatures that have them runs parallel to the logical distinction between the opaque 'selection-for' and the transparent 'selection of'. Accordingly, it's been a main point of our discussion that natural selection doesn't support inferences from 'creatures that have trait t were selected' to 'trait t was selected for in those creatures'. In practice, however, Darwinists draw such inferences all the time. (They call that 'reverse engineering', a polite term for post-hoc explanation.) The idea is to infer from an observed effect to a putative cause; as in 'lots of polar bears are white, so being white must have been good for fitness in the ecology in which they were selected.'

16. It's crucial that the idealizations are independently justified; otherwise 'all else being equal Fs cause Gs' collapses into 'Fs cause Gs except when they do not.'

17. Compare Sober (1993): 'Evolutionary biology has developed ... a system of models that describe the consequences for fitness of various traits' (p. 84). We think that is simply untrue; what it has developed is a system of models (typically post hoc) that explain why, de facto, certain traits affect fitness in certain circumstances. Sober offers the example 'in which heterozygote superiority explains a balanced polymorphism' (ibid.). But notice that this example concerns a genotypic trait, not a phenotypic trait. Since the relation between genotypic traits and their phenotypic expressions is generally quite indirect (see Part one), it's entirely possible that there should be laws about the effects on fitness of the first, even if there no laws about the effects on fitness of the second.

18. It bears emphasis that assuming the laws in question to be probabilistic wouldn't help. The problem is that 'it's probable that ...' is itself extensional and is thus unable to reconstruct the intensionality of 'selection-for ...' If being F is probably conducive to fitness, and F are G are coextensive, then being G is equally probably conducive to fitness; total gain, no yardage.

19. The field known as 'reintroduction biology' bears witness to the fact that, in many cases, planned human intervention that reintroduces species into an ecosystem where they once thrived often gives disappointing results. Maybe the quality of the soil has changed, maybe the populations of microorganisms are different, or whatever. This is to stress how hard it is to decide when two ecological niches count as being 'the same'. Some experts admit 'the poor success rate of reintroductions worldwide' (Armstrong and Seddon, 2008).

20. Protozoa not included; we gather that protozoa have recently become not-animals (much as Pluto has recently become a not-planet. *Sic transit* ...). We also haven't included viruses or

plants, all of which have been not-animals for a long time. If you do, the situation looks orders of magnitude worse.

21. Chapter 2 suggested (in the spirit of current 'evo-devo' theories of evolution) that much of the phylogentic structure that adaptationism attributes to environmental selection may in fact be an expression of endogenous variables. It's a virtue of this sort of proposal that, whereas it's very doubtful that there can be laws of selection (see text), there is no corresponding question about whether there are laws that connect endogenous features to their phenotypic expressions. The laws that specify genotype-to-phenotype relations are themselves instances of that kind.

22. We have slightly altered the details of Sober's example for purposes of exposition; but not in any way that matters to the discussion.

23. It is easy to imagine a photocell-operated sieve that would sort for colour, rather than for size. The final result would be the same, but in this case, knowing how it operates, we would conclude that it is sorting for colour, rather than for size.

24. Another way to put it: whether Sober's gadget selects for the stuff that gets to the bottom or for the stuff that's left on top depends on whether it's a sieve or a filter; but the difference between a sieve and a filter is entirely in the mind of the beholder.

25. The same applies, of course, to sorting-against. It's an occasional last refuge of neo-Darwinians, when forced to admit the importance of the sorts of considerations we've raised here and in Chapter 2, to claim that there still is an important role for natural selection: that of eliminating the (massively) unfit. But it's just a truism that, when there is competition between them, selection-for one kind of creature is selection-against some other kind. Selection-against is a by-product of selection-for (and vice versa).

26. Compare Dennett, 1995.

27. Here's an entirely typical instance of the kind; it's drawn from
a discussion of the evolution of neural specialization: 'The
flatworm's rudimentary division of neural labor was the first
step toward an avalanche of specialization that has given rise to
the complex neural systems of vertebrates … First, shortly after
vertebrates came on the scene, intercellular communication got
a whole lot better, with the evolution of glial cells … William
Richardson's research group has speculated that glial cells
evolved as modifications of motor neurons … They further
suggested that such glial cells could have immediately conveyed
a large adaptive advantage by making it possible for prey
to more rapidly escape their predators. It could, of course,
have been the other way around …' (Marcus, pp. 116–117).
There are, in this passage, hypotheses about the sequence of
phylogenetic alterations, and about the selectional advantages
that each step in the sequence might have conferred. And
although it is indeed a nomological generalization (physics
applied to nerve fibres) that myelinated fibres propagate
nervous impulses faster and more reliably, there's nothing in
Marcus's passage that sounds remotely like an attempt to frame
a covering law about ecology/phenotype interactions. It isn't
claimed, for example, that it's nomologically necessary that
neurons always (or generally) myelinate in the environment of
selection; or that phenotypes with myelinated neurons always
(or generally) win competitions with phenotypes that lack
them. What is suggested is that, as a matter of historical fact,
there were such competitions, that the myelinated phenotypes
won, and that it is intelligible (in retrospect) why they did so.
What's on offer as an explanation is a historical narrative, not
anything like a Hempelian deduction from generalizations to
their instances. But if, as would appear, the explanation makes
no appeal to laws, then a fortiori it makes no appeal to laws
of selection. Contrary to what the voice of exasperation says,
successful evolutionary explanations do not, in the general case,
depend on there being such laws.

28. Strictly speaking, of course, historical narratives aren't even of the right form to provide support for counterfactuals; you need quantified propositions to do that (e.g. propositions that quantify over both actual and possible states of affairs). Indeed, it's because quantified propositions can support counterfactuals and causal narratives cannot that there's a philosophical problem about induction: how can premises about what did happen justify conclusions about what always happens (and hence about what would happen if...)? Didn't Hume say something of that sort?

29. We've borrowed this example from Steven Schiffer, who uses it to argue that intentional explanations themselves aren't of the covering-law sort.

30. The issues here run exactly parallel to ones that are familiar from the philosophy of psychology. What's required in order to vindicate belief/desire psychology is that the laws that govern behaviour are laws about beliefs and desires as such; neurological laws, or quantum mechanical laws, would of course be lovely to have, but they wouldn't suffice to vindicate propositional attitude psychology. Compare Bunzl (2004, p. 7): '... just because historians don't explain events by reliance on laws of history, it does not follow that some explanations don't draw on laws of specific disciplines ... the assassination of Archduke Ferdinand involved a shot that was governed by the generalizations of physics'. Quite so; but since the laws of physics didn't apply to Ferdinand's assassination *qua* assassination, the explanations they afford abstract from precisely the aspects of the event that historians care about. What was interesting about Ferdinand was that he was assassinated, not that his mass was increased by a bullet's-worth. (For discussion, see Fodor, 'Special Sciences' and 'Reply to Kim'.) Bunzl remarks, at one point that 'implicit in every causal assertion, there is a set of counterfactual implications' (Bunzl, 2004, p. 13). But although that's true it is very misleading, since which counterfactuals are supported

depends on which laws the causal truths instantiate. That there are causal truths about historical events does not show – or even tend to show – that there are historical laws that subsume historical events.

31. It doesn't follow, of course, that the story about the mud (or the story about the Prussians, or even the story about Napoleon losing his touch) is just a 'just-so' story; i.e. that historical narratives ipso facto are ad hoc or ipso facto unconstrained. We think that Gould and his colleagues were a bit hard on adaptationists in this respect (although we also think their hearts were in the right place).

32. Dray (1964) acknowledges a 'grain of truth' in the view that '… if two historians make different selections out of what is known … there is no need to conclude that either of them writes a false account. Nor, for that matter, need we strictly speaking, regard them as contradicting one another. It is therefore somewhat misleading even to say that they offer different answers to the same question. Their answers are better regarded as providing "contributions" to the history of the subject in review.' Dray's book provides useful discussions of the sorts of differences between historical narratives and covering law (i.e. counterfactual supporting) explanations that we've been emphasizing here.

33. Nor, however, do the complications end with this. One of the reasons for their relative resistance to climate change (compared with the dinosaurs) was that the mammals had relatively small surface areas: that is, the mammals were smaller than the dinosaurs. So maybe the mammals won because of their size after all? It's precisely this kind of instability of the counterfactuals that suggests that there just aren't any laws that connect a creature's size, as such, to its success in competitions. (NB: it strongly suggests that; but, of course, it doesn't prove it.)

34. 'Everybody gets rich in contexts in which he accumulates riches' does not count (although it is, of course, perfectly

true). Likewise 'selection favours a creature that has found an ecological niche'.

35. Notice the immediate temptation to provide caveats ad hoc: 'Well, maybe there is a theory of how to get rich in twelfth-century Mongolia, and a different theory of how to get rich in the Wild West, and a still different theory of how to get rich in twenty-first-century Manhattan.' Maybe. On the other hand, maybe 'ways to get rich' just doesn't name a natural kind. To admit that would not be to deny that there's a difference between getting rich and not getting rich; or that that difference explains a lot about the differences between, say, Donald Trump and us.

36. We are, from time to time, accused of being covert Hempelians; but we're not. We see no reason to suppose that nomological explanations need be 'covering law' explanations in the sense of Hempel (Hempel, 1965); still less do we suppose that bona fide empirical explanations are ipso facto nomological (we take historical explanations to be instances to the contrary). Our point is just that if an explanation is to be other than post hoc, it must support relevant counterfactuals. Appealing to laws of selection might permit adaptationist theories to do so if there were laws of evolution for it to appeal to; but it appears that there aren't any.

Chapter 9

1. Skinner called behaviours that free-ride on reinforced responses 'superstitious' (Skinner, 1948). He had noticed that, on occasion, temporal proximity with the administration of a random reinforcer, just any piece of 'arbitrary' behaviour that a pigeon happened to be performing can become fixed (real instances include: turning counter-clockwise about the cage, thrusting its head into one of the upper corners of the cage, developing a pendulum motion of the head, etc.). But, of course, which behavior is considered 'arbitrary' and therefore

'superstitious' is entirely a question of what experimental design the experimenter has in mind; in particular, of what behaviour the experimenter intends to reinforce. The morals to draw are, first that it is considerably harder than Skinner supposed to develop an assiduously behaviourist psychology – one which really does prescind from intensional states and processes. No sooner than the pigeon's intentions go out one door, the experimenter's intentions come in the other. And, second, problems of free-riders are by no means proprietary to theories about evolution.

2. The notion of a 'level' of explanation gets thrown around with great self-confidence not just in biology and psychology but in many philosophical discussions of how scientific theories work. It is, however, extremely murky; perhaps a hundred philosophers could sort it out in a hundred years (fifty philosophers could probably do it twice as fast). For present purposes, however, we'll largely take it for granted. But it's worth remarking that, quite often, differences in levels of explanation correspond to differences in the size of the things that theories are about, the behaviour of the large things being explained by the behaviour of smaller ones that are their constituent parts. The laws that govern the former are said to 'emerge' (somehow, and whatever exactly that means) from the laws that govern the latter. That sciences are often arranged in this way has been clear to practically everybody at least since Lucretius; we suppose that something of the sort is likely to be true.

3. There is, in fact, considerable disagreement between biologists who view selection as a relation between ecological properties and phenotypic properties and biologists who view it as a relation between ecological properties and genetic or genomic properties. According to the former, selection is an organism–ecology relation that is mediated by genetic processes; according to the latter, it is constituted by genetic–ecological relations. But natural selection is a one-level theory in either view; the

argument is over the scale of the objects that enter into selection relations.

4. In a book that we think deserved greater attention and circulation than it received, the late political philosopher Robert Wesson wrote: 'Biologists, it seems, must do without a comprehensive theory of evolution, just as social scientists have to make do without a comprehensive theory of society' (Wesson, 1991, pp. xii–xiii). This applies, with bells on, to historians having to make do without a comprehensive theory of history.

5. This line of thought runs exactly parallel to Chomsky's argument that the fact that human languages do not vary at random strongly suggests the likelihood of endogenous constraints on what sorts of (first) languages are accessible to our species. It deserves to be stressed that his whole line of inquiry into language has been, and still is, fiercely contested by neo-Darwinian adaptationists. For radical critiques, see, among many, Arbib, 2005; Lieberman, 2006. For less extreme critiques, see Pinker and Bloom, 1990; Jackendoff, 2002; Jackendoff and Pinker, 2005.

REFERENCES

Abzhanov, A., Kuo, W. P., Hartmann, C., Grant, B. R., Grant, P. R. and Tabin, C. J., 'The calmodulin pathway and evolution of elongated beak morphology in Darwin's finches', *Nature*, vol. 442, 2006, pp. 563–67.

Agrawal, A., Eastman, Q. M. and Schatz, D. G., 'Implications of transposition mediated by $V(D)J$-recombination proteins RAG1 and RAG2 for origins of antigen-specific immunity', *Nature*, vol. 394, 1998, pp. 744–51.

Allis, C. D., Jenuwein, T., Reinberg, D. and Caparros, M.-L., *Epigenetics*, Cold Spring Harbor, Cold Spring Harbor Laboratory Press, 2006.

Amaral, P. P. and Mattick, J. S., 'Noncoding RNA in development', *Mammalian Genome*, vol. 19, 2008, pp. 454–92.

Amundson, R. A., 'EvoDevo as Cognitive Psychology', *Biological Theory*, vol. 1, 2006, pp. 10–11.

Anway, M. D. and Skinner, M. K., 'Epigenetic transgenerational actions of endocrine disruptors', *Endocrinology*, vol. 147 (6 Suppl.), 2006, pp. S43–S49.

Arbib, M. A., 'From monkey-like action recognition to human language: an evolutionary framework for neurolinguistics', *Behavioral and Brain Sciences*, vol. 28, 2005, pp. 105–67.

Ariew, A. and Lewontin, R. C., 'The confusions of fitness', *British Journal for the Philosophy of Science*, vol. 55, 2004, pp. 347–63.

Armstrong, D. P. and Seddon, P. J., 'Directions in reintroduction biology', *Trends in Ecology and Evolution*, vol. 23, 2008, pp. 20–25.

Baguna, J. and Garcia-Fernandez, J., 'Evo-devo: the long and winding road', *International Journal of Developmental Biology*, vol. 47, 2003, pp. 705–13.

Barrett, H. and Kurzban, R., 'Modularity in cognition: framing the debate', *Psychological Review*, vol. 113, 2006, pp. 628–47.

Barton, N. H., Briggs, D. E. G, Eisen, J. A., Goldstein, D. B. and Patel, N. H., *Evolution*, Cold Spring Harbor, NY, Cold Spring Harbor Laboratory Press (2007), Part III, Chapter 13 'Variation in DNA and proteins', 360.

Bejan, A. and Marden, J. H., 'Unifying constructal theory for scale effects in running, swimming and flying', *Journal of Experimental Biology*, vol. 209, 2006, pp. 238–48.

Ben-Tabou de Leon, S. and Davidson, E. H., 'Modeling the dynamics of transcriptional gene regulatory networks for animal development.' *Developmental Biology*, vol. 325, issue 2, 15 January 2009, 317–328.

Berglund, J., Pollard, K. S. and Webster, M. T., 'Hotspots of biased nucleotide substitutions in human genes', *PLoS Biology*, vol. 7, 2009, e26.

Block, N. and P. Kitcher, 'Misunderstanding Darwin' *Boston Review* vol. 35 (2), 2010, p. 29.

Blount, Z. D., Borland, C. Z. and Lensky, R. E., 'Historical contingency and the evolution of a key innovation in an experimental population of *Escherichia coli*, *Proceedings of the National Academy of Sciences USA*, vol. 105, 2008, pp. 7899–906.

Bois, P. R. J., 'Hypermutable minisatellites, a human affair?', *Genomics*, vol. 81, 2003, 349–55.

Boltzmann, L., 'On a thesis of Schopenhauer', translated and reprinted in B. McGuinness (ed.), *Theoretical Physics and Philosophical Problems*, Dordrecht, D. Reidel, 1974.

Boncinelli, E., *I Nostri Geni: la Natura Biologica dell'Uomo e le Frontiere della Ricerca*, Torino, Giulio Einaudi Editore, 1998.

Boncinelli, E., *Le Forme della Vita: l'Evoluzione e l'Origine dell'Uomo*, Torino, Giulio Einaudi Editore, 2000.

Bowler, P., *Evolution: the History of an Idea*, City, CA, University of California Press, 2003.

Bradshaw, A. D. and Hardwick, K., 'Evolution and stress – genotypic and phenotypic components', *Biological Journal of the Linnean Society*, vol. 37, 1989, pp. 137–55.

Brandon, R. N., 'Environment', in E. Fox Keller and E. A. Lloyd (eds), *Keywords in Evolutionary Biology*, Cambridge, MA, Harvard University Press, 1994, pp. 81–86.

Buchanan, M., 'Horizontal and vertical: the evolution of evolution.' *New Scientist* 2744, 2010, pp. 34–37.

Bunzl, M., 'Boas, Foucault, and the 'native anthropologist:' notes towards a neo-Boasian anthropology', *American Anthropologist*, vol. 106, 2004, pp. 435–42.

Buss, D., *The Handbook of Evolutionary Psychology*, Hoboken, NJ, Wiley, 2005.

Buss, D., 'The great struggles of life: Darwin and the emergence of evolutionary psychology', *American Psychologist*, vol. 64, 2009, p. 140–48.

Campbell, D., 'Evolutionary epistemology', in P. Schilpp (ed.), *The Philosophy of Karl Popper*, La Salle, IL, Open Court Press, 1974, pp. 413–63.

Campbell, D., Heyes, C. and Callebaut, W., 'Evolutionary epistemology bibliography', in W. Callebaut and R. Pinxten (eds), *Evolutionary Epistemology: a Multiparadigm Program*, Dordrecht, 1987, pp. 405–31.

Carroll, S. B., 'Homeotic genes and the evolution of arthropods and chordates', *Nature*, vol. 376, 1995, pp. 479–85.

Carroll, S. B., *Endless Forms Most Beautiful: the New Science of Evo Devo and the Making of the Animal Kingdom*, New York, W. W. Norton & Company, 2005.

Carroll, S. B., *The Making of the Fittest: DNA and the Ultimate Forensic Record of Evolution*. New York, W. W. Norton & Company, 2006.

Cartwright, J., 'News: Fibonacci spirals in nature could be stress-related', 2007, available online at physicsworld.com: http://physicsworld.com/cws/article/news/27722 [accessed August 2009].

Cherniak, C., 'Brain wiring optimization and non-genomic nativism' in M. Piattelli-Palmarini, J. Uriagereka and P. Salaburu (eds), *Of Minds and Language: a Dialogue with Noam Chomsky in the Basque Country*, Oxford, Oxford University Press, 2009, pp. 108–19.

Cherniak, C., Changizi, M. and Kang, D., 'Large-scale optimization of neuron arbors', *Physical Review E, Statistical Physics, Plasmas, Fluids, and Related Interdisciplinary Topics*, vol. 59, 1999, p. 6001.

Cherniak, C., Mokhtarzada, Z., Rodriguez-Esteban, R. and Changizi, K., 'Global optimization of cerebral cortex layout', *Proceedings of the National Academy of Sciences USA*, vol. 101, 2004, pp. 1081–86.

Chomsky, N., 'A review of B. F. Skinner's *Verbal Behavior*', *Language*, vol. 35, 1959, pp. 26–58. Reprinted in N. Block (ed.), *Readings in the Philosophy of Psychology*, Volume 1, The Language and Thought Series, Cambridge, MA, Harvard University Press, 1980.

Chomsky, N., *Language and Mind*, New York, Harcourt, Brace & Jovanovich, 1972.

Chomsky, N., *Lectures on Government and Binding: the Pisa Lectures*. Dordrecht, Foris Publications, 1981.

Chomsky, N., 'Opening remarks', in M. Piattelli-Palmarini, J. Uriagereka and P. Salaburu (eds), *Of Minds and Language: a Dialogue with Noam Chomsky in the Basque Country*, Oxford, Oxford University Press, 2009, pp. 13–43.

Churchland, P., 'Eliminative materialism and the propositional attitudes', *Journal of Philosophy*, vol. 78, 1981, pp. 67–90.

Cokely, E. T. and Feltz, A., 'Adaptive variation in judgment and philosophical intuition', *Consciousness and Cognition*, vol. 18, 2009, pp. 356–58.

Coller, H. A. and Kruglyak, L., 'It's the sequence, stupid!', *Science*, vol. 322, 2008, pp. 380–81.

Cosmides, L. and Tooby, J., 'Better than rational: evolutionary psychology and the invisible hand', *American Economic Review*, vol. 84, 1994, pp. 327–32.

Cowperthwaite, M. C., Economo, E. P., Harcombe, W. R., Miller, E. L. and Meyers, L. A., 'The ascent of the abundant: how mutational networks constrain evolution', *PLoS Computational Biology*, vol. 4, 2008, e1000110.

Coyne, J. A., 'Comment on "gene regulatory networks and the evolution of animal body plans"', *Science*, vol. 313, 2006, p. 761.

Coyne, J. A., *Why Evolution is True*, New York, Viking Penguin, 2009.

Coyne, J. A., 'The improbability pump', *The Nation*, 10 May, 2010.

Daly, M. and Wilson, M., *Homicide*, Aldine de Gruyter, 1988.

Daly, M. and Wilson, M., 'Is the 'Cinderella Effect' controversial?', in C. Crawford and D. Krebs (eds), *Foundations of Evolutionary Psychology*, New York, Psychology Press, 2008, pp. 383–400.

Darwin, C., *On the Origin of Species by Natural Selection*, London, Murray, 1859, p. 458.

Davidson, E. H., *The Regulatory Genome: Gene Regulatory Networks in Development and Evolution*, London, Elsevier Academic Press, 2006.

Davidson, E. H. and Erwin, D. H., 'Gene regulatory networks and the evolution of animal body plans', *Science*, vol. 311, 2006, 796–800.

Dawkins, R., *The Blind Watchmaker: Why the Evidence of Evolution Reveals a Universe Without Design*, New York, W. W. Norton & Company, 1986.

Dechaume-Moncharmont, F.-X., Dornhaus, A., Houston, A. I., McNamara, J. M., Collins, E. J. and Franks, N. R., 'The hidden cost of information in collective foraging', *Proceedings of the*

Royal Society of London. Series B. Biological Sciences, vol. 272, 2005, 1689–95.

Dehaene, S., Changeux, J. P. and Nadal, J. P., 'Neural networks that learn temporal sequences by selection', *Proceedings of the National Academy of Sciences USA* vol. 84, 1987, pp. 2727–31.

De Leon, B. S. B. T. and Davidson, E. H., 'Review: modeling the dynamics of transcriptional gene regulatory networks for animal development', *Developmental Biology*, vol. 325, 2009, pp. 317–28.

Dennett, D., *Darwin's Dangerous Idea: Evolution and the Meanings of Life*, New York, Simon & Schuster, 1995.

Dennett, D., 'Response to Fodor "Why pigs don't have wings"' [letter], *London Review of Books*, vol. 29, 2007.

Dennett, D., 'Fun and games in fantasyland', *Mind & Language*, vol. 23, 2008, pp. 25–31.

Dewey, J., *The Influence of Darwin on Philosophy and Other Essays in Contemporary Thought*, New York, Henry Holt and Company, 1910.

Dial, K., Jackson, B. and Segre, P., 'A fundamental avian wing-stroke provides a new perspective on the evolution of flight', *Nature*, vol. 451, 2008, pp. 985–89.

Diamond, J., Gilpin, M. and Mayr, E., 'Species-distance relation for birds of the Solomon Archipelago, and the paradox of the great speciators', *Proceedings of the National Academy of Sciences USA*, vol. 73, 1976, pp. 2160–64.

Dobson, C. M., 'Protein folding and misfolding', *Nature*, vol. 426, 2003, pp. 884–90.

Dobzhansky, T., 'Nothing in biology makes sense except in the light of evolution', *American Biology Teacher*, vol. 35, 1973, pp. 125–29.

Doolittle, W. F., 'Phylogenetic classification and the universal tree', *Science*, vol. 284, 1999, pp. 2124–28.

Douady, S. and Couder, Y., 'Phyllotaxis as a physical self-organized growth process', *Physical Review Letters*, vol. 68, 1992, pp. 2098–101.

Dover, G. A., 'A molecular drive through evolution', *BioScience*, vol. 32, 1982a, pp. 526–33.

Dover, G. A., 'Molecular drive: a cohesive mode of species evolution', *Nature*, vol. 299, 1982b, 111–17.

Dover, G. A., *Dear Mr. Darwin: Letters on the Evolution of Life and Human Nature*, London, Orion Publishing, 2001.

Dover, G. A., 'Darwin and the idea of natural selection', *Encyclopedia of Life Sciences*, London, John Wiley and Sons Ltd, 2006.Dray, W., *Philosophy of History*, Englewood Cliffs, NJ, Prentice-Hall, 1964.

Duret, L., 'Mutation patterns in the human genome: more variable than expected', *PLoS Biology*, vol. 7, 2009, e28.

Edelman, G., *Neural Darwinism: the Theory of Neuronal Group Selection*, New York, Basic Books, 1987.

Eder, W., Klimecki, L., Yu, E., von Mutius, J., Riedler, C., Braun-Fahrlander, D. *et al.*, 'Toll-like receptor 2 as a major gene for asthma in children of European farmers', *Journal of Allergy and Clinical Immunology*, vol. 113, 2004, pp. 482–88.

Eldredge, N., *Reinventing Darwin: the Great Evolutionary Debate*, Phoenix, 1996.

Ellis, B., *The Philosophy of Nature: a Guide to the New Essentialism*, Montreal, McGill-Queens University Press, 2002.

Engel, M. and Grimaldi, D., 'New light shed on the oldest insect', *Nature*, vol. 427, 2004, pp. 627–30.

Erwin, D. H., 'Wonderful ediacarans, wonderful cnidarians?', *Evolution & Development*, vol. 10, 2008, pp. 263–64.

Erwin, D. H. and Davidson, E. H., 'Response to comment on "gene regulatory networks and the evolution of animal body plans"', *Science*, vol. 313, 2006, p. 761.

Ewens, W. J., 'Commentary: On Haldane's "defense of beanbag genetics"', *International Journal of Epidemiology*, vol. 37, 2008, pp. 447–51.

Felix, M.-A. and Wagner, A., 'Robustness and evolution: concepts, insights and challenges from a developmental model system', *Heredity*, vol. 100, 2008, pp. 132–40.

Feuerhahn, S. and Egly, J.-M., 'Tools to study DNA repair: what's in the box?', *Trends in Genetics*, vol. 24, 2008, pp. 467–74.

Filipowicz, W., Bhattacharyya, S. N. and Sonenberg, N., 'Mechanisms of post-transcriptional regulation by microRNAs: are the answers in sight?', *Nature Reviews Genetics*, vol. 9, 2008, pp. 102–14.

Fisher, S. E., 'Tangled webs: tracing the connections between genes and cognition', *Cognition*, vol. 101, 2006, pp. 270–97.

Fodor, J., *A Theory of Content and Other Essays*, Cambridge, MA, MIT Press, 1990.

Fodor, J., *Lo2: The Language of Thought Revisited*, New York, Oxford, 2008.

Fouad, K., Libersat, F. and Rathmayer, W., 'The venom of the cockroach-hunting wasp *Ampulex compressa* changes motor thresholds: a novel tool for studying the neural control of arousal', *Zoology*, vol. 98, 1994, pp. 23–34.

Fraga, M. F., Ballestar, E., Paz, M. F., Ropero, S., Setien, F., Ballestar, M. L. *et al.*, 'Epigenetic differences arise during the lifetime of monozygotic twins', *Proceedings of the National Academy of Sciences USA*, vol. 102, 2005, pp. 10604–09.

Fusco, G., 'How many processes are responsible for phenotypic evolution?', *Evolution & Development*, vol. 3, 2001, pp. 279–86.

Futuyma, D. J., 'Two critics without a clue', *Science*, vol. 328, 7 May, 2010, pp. 692–3.

Gallistel, C. R., 'The replacement of general-purpose learning models with adaptively specialized learning modules', in M. S. Gazzaniga (ed.), *The New Cognitive Neuroscience*, 2nd edn, Cambridge, MA, Bradford Books/MIT Press, 2000, pp. 1179–91.

Gallistel, C. R., 'Frequency, contingency and the information processing theory of conditioning', in P. Sedlmeier and T. Betsch (eds), *Frequency Processing and Cognition*, Oxford, Oxford University Press, 2002, pp. 153–71.

Galtier, N., Duret, L., Glémin, S. and Ranwez, V., 'Gc-biased gene conversion promotes the fixation of deleterious amino acid changes in primates', *Trends in Genetics*, vol. 25, 2009, pp. 1–5.

Gauthier, A. S., Furstoss, O., Araki, T., Chan, R., Neel, B. G., Kaplan, D. R. *et al.*, 'Control of cns cell-fate decisions by shp-2 and its dysregulation in noonan syndrome', *Neuron*, vol. 54, 2007, pp. 245–62.

Gibson, G., 'Systems biology: the origins of stability', *Science*, vol. 310, 2005, p. 237.

Godfrey-Smith, P., 'It Got Eaten', review in *London Review of Books*, vol. 32 (13), 2010, pp. 29–30. (http://www.lrb.co.uk/v32/n13/peter-godfrey-smith/it-got-eaten)

Goldstein, D. B., 'Common Genetic Variation and Human Traits.' *New England Journal of Medicine*, vol. 360 (17), 23 April, 2009, pp. 1696–8.

González, J., Lenkov, K., Lipatov, M., Macpherson, J. M. and Petrov, D. A., 'High rate of recent transposable element–induced adaptation in *Drosophila melanogaster*', *PloS Biology*, vol. 6, 2008, e251.

Gopnik, A., 'Rewriting nature', *The New Yorker*, vol. 82, 2006, pp. 52–59.

Gould, S. J., *Quando i Cavalli Avevano le Dita*, Milan, Feltrinelli, 1989.

Gould, S. J. and Lewontin, R. C., 'The spandrels of San Marco and the Panglossian paradigm: a critique of the adaptationist programme', *Proceedings of the Royal Society of London. Series B. Biological Sciences*, vol. 205, 1979, pp. 581–98.

Hall, B. G., 'Experimental evolution of a new enzymatic function. II. Evolution of multiple functions for ebg enzyme in *E. coli*', *Genetics*, vol. 89, 1978, pp. 453–65.

Hall, B. G., 'Changes in the substrate specificities of an enzyme during directed evolution of new functions', *Biochemistry*, vol. 20, 1981, 4042–49.

Hempel, C. G., *Aspects of Scientific Explanation and Other Essays in the Philosophy of Science*, New York, Free Press, 1965.

Heo, M., Kang, L. and Shakhnovich, E. I., 'Emergence of species in evolutionary "simulated annealing"', *Proceedings of the National Academy of Sciences USA*, vol. 106, 2009, 1869–74.

Hiom, K., Melek, M. and Gellert, M., 'DNA transposition by the rag1 and rag2 proteins: a possible source of oncogenic translocations', *Cell*, vol. 94, 1998, pp. 463–70.

Hodgkinson, A., Ladoukakis, E. and Eyre-Walker, A., 'Cryptic variation in the human mutation rate', *PLoS Biology*, vol. 7, 2009, e1000027.

Hoenigsberg, H., 'The future of selection: individuality, the twin legacies of Lamarck & Darwin', *Genetics and Molecular Research*, vol. 1, 2002, 39–50.

Hurst, L. D., 'A positive becomes a negative', *Nature*, vol. 457, 2009, pp. 543–44.

Itzkovitz, S. and Alon, U., 'The genetic code is nearly optimal for allowing arbitrary additional information within protein-coding sequences', *Genome Research*, vol. 17, 2007, pp. 405–12.

Jackendoff, R., *Foundations of Language: Brain, Meaning, Grammar, Evolution*, New York, Oxford University Press, 2002.

Jackendoff, R. and Pinker, S., 'The nature of the language faculty and its implications for evolution of language (reply to Fitch, Hauser and Chomsky)', *Cognition*, vol. 97, 2005, pp. 211–25.

Jirtle, R. L. and Skinner, M. K., 'Environmental epigenomics and disease susceptibility', *Nature Reviews Genetics*, vol. 8, 2007, pp. 253–62.

Jockusch, E. L. and Ober, K. A., 'Hypothesis testing in evolutionary developmental biology: a case study from insect wings', *Journal of Heredity*, vol. 95, 2004, pp. 382–96.

Kaati, G., Bygren, L. D., and Edvinsson, S., 'Cardiovascular and diabetes mortality determined by nutrition during parents' and grandparents' slow growth period', *European Journal of Human Genetics*, vol. 10, 2002, pp. 682–88.

Kaminsky, Z. A., Tang, T., Wang, S.-C., Ptak, C., Oh, G. H. T., Wong, A. H. C. *et al.*, 'DNA methylation profiles in monozygotic and dizygotic twins', *Nature Genetics*, vol. 41, 2009, pp. 240–45.

Kauffman, S., 'Developmental logic and its evolution', *BioEssays*, vol. 6, 1987, pp. 82–87.

Kauffman, S. A., *The Origins of Order: Self-Organization and Selection in Evolution*, Oxford, Oxford University Press, 1993.

Kendler, H., 'What is learned? A theoretical blind alley', *Psychological Review*, vol. 59, 1952, pp. 269–77.

Kingsolver, J. and Koehl, M., 'Aerodynamics, thermoregulation, and the evolution of insect wings: differential scaling and evolutionary change', *Evolution*, vol. 39, 1985, pp. 488–504.

Kioussis, D., 'Gene regulation: kissing chromosomes', *Nature*, vol. 435, 2005, pp. 579–80.

Kirschner, M. W. and Gerhart, J. C., *The Plausibility of Life: Resolving Darwin's Dilemma*, New Haven, CT, Yale University Press, 2005.

Kitano, H., 'Biological robustness', *Nature Reviews Genetics*, vol. 5, 2004, pp. 826–37.

Koonin, E. V., 'The "Origin" at 150: is a new evolutionary synthesis in sight?' *Trends in Genetics* vol. 25 (11), 2009, pp. 473–5.

Koonin, E. V., 'Darwinian evolution in the light of genomics.' *Nucleic Acids Research* vol. 37 (4), 2009, pp. 1011–34.

Laken, S. J., Petersen, G. M., Gruber, S. B., Oddoux, C., Ostrer, H., Giardiello, F. M., *et al.*, 'Familial colorectal cancer in Ashkenazim due to a hypermutable tract in APC', *Nature Genetics*, vol. 17, 1997, pp. 79–83.

Leiber, J., 'Turing and the fragility and insubstantiality of evolutionary explanations: a puzzle about the unity of Alan Turing's work with some larger implications', *Philosophical Psychology*, vol. 14, 2001, pp. 83–94.

Levit, G. S., Hossfeld, U. and Olsson, L., 'From the "Modern synthesis" to cybernetics: Ivan Ivanovich Schmalhausen (1884–1963) and his research program for a synthesis of evolutionary and developmental biology', *Journal of Experimental Zoology. Part B. Molecular and Developmental Evolution*, vol. 306, 2006, pp. 89–106.

Lewis, E. B., Nüsslein-Volhard, C. and Wieschaus, E., in N. Ringertz (ed.), *Nobel Lectures, Physiology or Medicine 1991–1995*, Singapore, World Scientific Publishing Co., 1997.

Lewontin, R. C., 'The evolution of cognition: questions we will never answer', in D. Scarborough and S. Sternberg (eds), *An Invitation to Cognitive Science: Vol. 4 Methods, Models and Conceptual Issues*, Cambridge, MA, The MIT Press, 1998, pp. 107–32.

Lewontin, R. C., *The Triple Helix: Gene, Organism and Environment*, Cambridge, MA, Harvard University Press, 2000.

Lewontin, R. C., Paul, D. B., Beatty, J. and Krimbas, C. B., 'Interview with R. C. Lewontin', in R. S. Singh, C. B. Krimbas, D. B. Paul and J. Beatty (eds), *Thinking about Evolution: Historical, Philosophical, and Political Perspectives (Essays in Honor of Richard Lewontin)*, Vol. 2, Cambridge, UK, Cambridge University Press, 2001, pp. 22–63.

Li, C., Ji, A. and Cao, Z., 'Stressed Fibonacci spiral patterns of definite chirality', *Applied Physics Letters*, vol. 90, 2007, 164102.

Li, G., Ji, P., Sun, L. Y. and Lee, W. B., 'Modeling and simulation of supply network evolution based on complex adaptive system and fitness landscape', *Computers and Industrial Engineering*, vol. 56, 2008, pp. 839–53.

Libersat, F., 'Wasp uses venom cocktail to manipulate the behavior of its cockroach prey', *Journal of Comparative Physiology A*, vol. 189, 2003, pp. 497–508.

Lickliter, R. and Honeycutt, H., 'Developmental dynamics and contemporary evolutionary psychology: status quo or irreconcilable views? Replies to critics', *Psychological Bulletin*, vol. 129, 2003a, pp. 866–72.

Lickliter, R. and Honeycutt, H., 'Developmental dynamics: toward a biologically plausible evolutionary psychology', *Psychological Bulletin*, vol. 129, 2003b, pp. 819–35.

Lieberman, P., *Toward an Evolutionary Biology of Language*, Cambridge, MA, Harvard University Press, 2006.

Lin, L., Shen, S., Tye, A., Cai, J. J. *et al.* 'Diverse splicing patterns of exonized Alu elements in human tissues', *PLoS Genetics*, vol. 4, 2008, e1000225.

Lotka, A. J., *Elements of Mathematical Biology*, New York, Dover, 1956. Originally published as Lotka, A. J. (1925) *Elements of Physical Biology*, Baltimore, MD, Williams and Wilkins Company.

MacFarlane, D. A., 'The role of kinesthesis in maze learning', *University of California Publications in Psychology*, vol. 4, 1930, pp. 277–305.

Machery, E. and Barrett, H., 'Essay review: debunking adapting minds', *Philosophy of Science*, vol. 73, 2006, pp. 232–46.

Malecek, K., Lee, V., Feng, W., Huang, J. L., Flajnik, M. F., Ohta, Y. *et al.*, 'Immunoglobulin heavy chain exclusion in the shark', *PLoS Biology*, vol. 6, 2008, e157.

Malicki, J., Cianetti, L. C., Peschle, C. and McGinnis, W., 'A human HOX4B regulatory element provides head-specific expression in *Drosophila* embryos', *Nature*, vol. 358, 1992, pp. 345–47.

Mallam, A. L., Morris, E. R. and Jackson, S. E., 'Exploring knotting mechanisms in protein folding', *Proceedings of the National Academy of Sciences USA*, vol. 105, 2008, pp. 18740–45.

Manolio, T. A., F. S. Collins, N. J. Cox, D. B. Goldstein, L. A. Hindorff, D. J. Hunter, M. I. McCarthy, E. M. Ramos, L. R. Cardon, A. Chakravarti, J. H. Cho, A. E. Guttmacher, A. Kong, L. Kruglyak, E. Mardis, C. N. Rotimi, M. Slatkin, D. Valle, A. S. Whittemore, M. Boehnke, A. G. Clark, E. E. Eichler, G. Gibson, J. L. Haines, T. F. C. Mackay, S. A. McCarroll and P. M. Visscher, 'Finding the missing heritability of complex diseases.' *Nature* vol. 461, 8 October, 2009, pp. 747–53.

Marcus, Gary F. *The Birth of the Mind: how a tiny number of genes create the complexities of human thought*, New York, Basic Books, 2004.

Marcus, G., *The Birth of the Mind*, New York, Basic Books, 2004.

Margulis, L. and L. J. Chapman, *Kingdoms & Domains: An Illustrated Guide to the Phyla of Life on Earth*, San Diego, London, Elsevier, 2010.

Martinez, F. D., 'Gene–environment interactions in asthma: with apologies to William of Ockham', *Proceedings of the American Thoracic Society*, vol. 4, 2007, pp. 26–31.

Mattick, J. S., 'The functional genomics of noncoding RNA', *Science*, vol. 309, 2005, pp. 1527–28.

Mattick, J. S. and Mehler, M. F., 'RNA editing, DNA recoding and the evolution of human cognition', *Trends in Neurosciences*, vol. 31, 2008, pp. 227–33.

Maynard Smith, J. and Savage, R. J. G., 'Some adaptations in mammals', *Biological Journal of the Linnean Society*, vol. 42, 1956, pp. 603–22.

Maynard Smith, J., Burian, R., Kauffman, S., Alberch, P., Campbell, J., Goodwin, B., *et al.*, 'Developmental constraints and evolution: a perspective from the mountain lake conference on development and evolution', *Quarterly Review of Biology*, vol. 60, 1985, pp. 265–87.

Mayr, E., *Animal Species and Evolution*, Cambridge, MA, Harvard University Press, 1963.

Mayr, E., *What Evolution Is*, New York, Basic Books, 2001.

McGhee, G., *The Geometry of Evolution: Adaptive Landscapes and Theoretical Morphospaces*, Cambridge, UK, Cambridge University Press, 2007.

McKinney, F. K. and McGhee, G. R., 'Evolution of erect helical colony form in the bryozoa: phylogenetic, functional, and ecological factors', *Biological Journal of the Linnean Society*, vol. 80, 2003, pp. 235–60.

McKinney, M. L. and Gittelman, J. L., 'Ontogeny and philogeny: tinkering with covariation in life history, morphology and behaviour', in K. J. McNamara (ed.), *Evolutionary Change and Heterochrony*, Chichester, Wiley, 1995, pp. 21–47.

Medeiros, D. P., 'Optimal growth in phrase structure', *Biolinguistics*, vol. 2, 2008, pp. 152–95.

Meister, P., Poidevin, M., Francesconi, S., Tratner, I., Zarzov, P. and Baldacci, G., 'Nuclear factories for signalling and repairing

DNA double strand breaks in living fission yeast', *Nucleic Acids Research*, vol. 31, 2003, pp. 5064–5073.

Michod, R. E., *Darwinian Dynamics: Evolutionary Transitions in Fitness and Individuality*, Princeton, NJ, Princeton University Press, 1999.

Mill, J. S., *A System of Logic: Ratiocinative and Inductive*, New York, Harper and Brothers, 1846, p. 469.

Millikan, R., 'The language–thought partnership a bird's eye view', *Language and Communication*, vol. 21, 2001, pp. 157–66.

Millikan, R., *Varieties of Meaning: the 2002 Jean-Nicod Lectures*, Cambridge, MA, MIT Press, 2004.

Moyle, R. G., Filardi, C. E., Smith, C. E. and Diamond, J., 'Explosive Pleistocene diversification and hemispheric expansion of a "great speciator"', *Proceedings of the National Academy of Sciences USA*, vol. 106, 2009, pp. 1863–68.

Murphy, D. and Stich, S., 'Darwin in the madhouse: evolutionary psychology and the classification of mental disorders', in Peter Carruthers and Andrew Chamberlain (eds), *Evolution and the Human Mind: Modularity, Language and Meta-Cognition*, Cambridge, UK, Cambridge University Press, 2000, pp. 62–92, available online at: http://www.philosophy.uconn.edu/department/millikan/ratam.pdf [accessed August 2009], pp. 1–34.

Newman, S. A. and Bhat, R., 'Dynamical patterning modules: physico-genetic determinants of morphological development and evolution', *Physical Biology*, vol. 5, 2008, 15008.

Nijhout, H. F., 'The nature of robustness in development', *BioEssays*, vol. 24, 2002, pp. 553–63.

Noblin, X., Mahadevan, L., Coomaraswamy, I., Weitz, D., Holbrook, N. and Zwieniecki, M., 'Optimal vein density in artificial and real leaves', *Proceedings of the National Academy of Sciences USA*, vol. 105, 2008, pp. 9140–9144.

Okasha, S., Review of Fodor and Piattelli-Palmarini's *What Darwin GotWrong*, *Times Literary Supplement*, 26 March, 2010.

Oliveri, P. and Davidson, E. H., 'Development: built to run, not fail', *Science*, vol. 315, 2007, pp. 1510–11.

Papineau, D., Review of Fodor and Piattelli-Palmarini's *What Darwin Got Wrong, Prospect*, vol. 168, pp. 83–4, 2010 http://www.kcl.ac.uk/content/1/c6/04/17/80/FodorDarwinrevProspect.doc.

Peirce, C., 'The fixation of belief', *Popular Science Monthly*, vol. 12, 1877, pp. 1–15.

Pembrey, M. E., 'Time to take epigenetic inheritance seriously', *European Journal of Human Genetics*, vol. 10, 2002, pp. 669–71.

Pennisi, E., 'Evolutionary biology. Evo-devo enthusiasts get down to details', *Science*, vol. 298, 2002, pp. 953–55.

Peters, J. and Robson, J. E., 'Imprinted non-coding RNAs', *Mammalian Genome*, vol. 19, 2008, pp. 493–502.

Piattelli-Palmarini, M. (ed.), *Language and Learning: the Debate between Jean Piaget and Noam Chomsky*, Cambridge, MA, Harvard University Press, 1980.

Piattelli-Palmarini, M., 'Novel tools at the service of old ideas', *Biolinguistics*, vol. 2, 2008, pp. 185–94.

Piattelli-Palmarini, M. and Uriagereka, J., 'Still a bridge too far? Biolinguistic questions for grounding language on brains', *Physics of Life Reviews*, vol. 5, 2008, pp. 207–24.

Pigliucci, M. and Kaplan, J., *Making Sense of Evolution: the Conceptual Foundations of Evolutionary Biology*, Chicago, IL, The University of Chicago Press, 2006.

Pinker, S., *How the Mind Works*, New York, W. W. Norton & Company, 1997.

Pinker, S. and Bloom, P., 'Natural language and natural selection', *Behavioral and Brain Sciences*, vol. 13, 1990, pp. 707–84.

Poelwijk, F. J., Kiviet, D. J., Weinreich, D. M. and Tans, S. J., 'Empirical fitness landscapes reveal accessible evolutionary paths', *Nature*, vol. 445, 2007, pp. 383–86.

Popper, K., 'The rationality of scientific revolutions', in R. Harre (ed.), *Problems of Scientific Revolution*, Oxford, Clarendon Press, 1975, pp. 72–101.

Popper, K., 'Unended quest: an intellectual biography', London, Fontana, 1976.

Prabhakar, S., Visel, A., Akiyama, J. A., Shoukry, M., Lewis, K. D., Holt, A. *et al.*, 'Human-specific gain of function in a developmental enhancer', *Science*, vol. 321, 2008, pp. 1346–50.

Pray, L. A., 'Epigenetics: genome, meet your environment', *The Scientist*, 18, 2004, pp. 1–10.

Prigogine, I., 'Time, structure and fluctuations', in S. Forsén (ed.), *Nobel Lectures, Chemistry 1971–1980*, Singapore, World Scientific Publishing Co., 1993.

Provine, W. B., 'Progress in evolution and meaning in life'. In *Evolutionary Progress*. M. H. Nitecki (ed.), Chicago, IL, University of Chicago Press, 1988, pp. 49–74.

Provine, W. B., *The Origins of Theoretical Population Genetics*, Chicago, IL, University of Chicago Press, 1971 (republished 2001).

Pigliucci, M. (2009a) 'An extended synthesis for evolutionary biology?' *Annals of the New York Academy of Sciences* 1168 (*The Year in Evolutionary Biology 2009*), pp. 218–28.

Pigliucci, M. (2009b) 'Down with natural selection?' *Perspectives in Biology and Medicine* vol. 52 (Winter), pp. 134–40.

Queltsch, C., Sangster, T. A. and Lindquist, S., 'Hsp90 as a capacitor of phenotypic variation', *Nature*, vol. 417, 2002, pp. 618–24.

Quine, W. V. O., 'Natural kinds', in N. Rescher (ed.) *Essays in Honor of Carl G. Hempel*, Dordrecht, Reidel, 1969, pp. 5–23. Reprinted in Quine *Ontological Relativity and Other Essays*, Irvington, NY, Columbia University Press, 1997.

Quinn, A. E., Georges, A., Sarre, S. D., Guarino, F., Ezaz, T. and Graves, J. A. M., 'Temperature sex reversal implies sex gene dosage in a reptile', *Science*, vol. 316, 2007, p. 411.

Radnitzky, G., Bartley, W. and Popper, K., *Evolutionary Epistemology, Rationality, and the Sociology of Knowledge*, Chicago, IL, Open Court Publishing, 1987.

Rakic, P., 'Confusing cortical columns', *Proceedings of the National Academy of Sciences USA*, vol. 105, pp. 12099–100.

Rastogy, N. and Pandey, M.,'Statistical analysis of geographical variability in 16 ecotypes of Indian Hydra', *Evolutionary Biology*, vol. 6, 1992, pp. 195–204.

Raup, D. M., 'Geometric analysis of shell coiling: general problems', *Journal of Paleontology*, vol. 40, 1966, pp. 1178–90.

Raup, D. M., 'Geometric analysis of shell coiling: coiling in ammonoids', *Journal of Paleontology*, vol. 41, 1967, pp. 43–65.

Rendel, J. M., 'Genetic control of a developmental process', in R. C. Lewontin (ed.), *Population Biology and Evolution*, Syracuse, NY, Syracuse University Press, 1968, pp. 47–66.

Rendel, J. M., 'Model relating gene replicas and gene repression to phenotypic expression and variability', *Proceedings of the National Academy of Sciences USA*, vol. 64, 1969, pp. 578–83.

Restifo, L. L., 'Mental retardation genes in *Drosophila*: new approaches to understanding and treating developmental brain disorders', *Mental Retardation and Developmental Disabilities Research Reviews*, vol. 11, 2005, pp. 286–94.

Rice, S. H., *Evolutionary Theory: Mathematical and Conceptual Foundations*, Sunderland, MA, Sinauer Associates, 2004.

Richards, R., *Darwin and the Emergence of Evolutionary Theories of Mind and Behavior*, Chicago, IL, University of Chicago Press, 1987.

Rizzi, L., *Relativized Minimality*, Linguistic Inquiry Monograph Series, Cambridge, MA, MIT Press, 1989.

Ronshaugen, M., McGinnis, N. and McGinnis, W., 'Hox protein mutation and macroevolution of the insect body plan', *Nature*, vol. 415, 2002, pp. 914–17.

Rueber, L. and Adams, D. C., 'Evolutionary convergence of body shape and trophic morphology in cichlids from Lake Tanganyika', *Journal of Evolutionary Biology*, vol. 14, 2001, 325–32.

Rutherford, S. L. and Henicoff, S., 'Quantitative epigenetics', *Nature Genetics*, vol. 33, 2003, pp. 6–8.

Rutherford, S. L. and Lindquist, S., 'Hsp90 as a capacitor for morphological evolution', *Nature*, vol. 396, 1998, pp. 336–42.

Sangster, T. A., Salathia, N., Lee, H. N., Watabnabe, E., Schellenberg, K., Morneau, K. *et al.*, 'Hsp90-buffered genetic variation is common in *Arabidopsis thaliana*', *Proceedings of the National Academy of Sciences USA*, vol. 105, 2008, pp. 2969–74.

Saunders, P. T., *An Introduction to Catastrophe Theory*, Cambridge, UK, Cambridge University Press, 1980.

Saunders, P. T. (ed.), *Morphogenesis: Collected Works of A. M. Turing*, Volume 3, Amsterdam, North-Holland, 1992.

Schank, J. J. and Wimsatt, W. C., 'Evolvability: adaptation and modularity', in R. S. Singh, C. B. Krimbas, D. B. Paul and J. Beatty (eds), *Thinking about Evolution: Historical, Philosophical, and Political Perspectives (Essays in Honor of Richard Lewontin)*, Volume 2, Cambridge, UK, Cambridge University Press, 2001, pp. 322–35.

Schlosser, G., 'The role of modules in development and evolution', in G. Schlosser and G. P. Wagner (eds), *Modularity in Development and Evolution*, Chicago, IL, University of Chicago Press, 2004, pp. 519–82.

Schlosser, G. and Wagner, G., *Modularity in Development and Evolution*, Chicago, IL, University of Chicago Press, 2004.

Schmucker, D. and Chen, B., 'Dscam and DSCAM: complex genes in simple animals, complex animals yet simple genes', *Genes and Development*, vol. 23, 2009, pp. 147–156.

Seager, William, 'The "Intrinsic Nature" Argument for Panpsychism' *Journal of Consciousness Studies*, Vol. 13, No. 10–11, 2006, 129–45.

Shen, B., Dong, L., Xiao, S. and Kowalewski, M., 'The Avalon explosion: evolution of Ediacara morphospace', *Science*, vol. 319, 2008, pp. 81–84.

Shen, H. M. and Storb, U., 'Activation-induced cytidine deaminase (AID) can target both DNA strands when the DNA is supercoiled', *Proceedings of the National Academy of Sciences USA*, vol. 101, 2004, pp. 12997–13002.

Sherman, M., 'Universal genome in the origin of metazoa: thoughts about evolution', *Cell Cycle*, vol. 6, 2007, 1873–77.

Silander, O. K., Tenaillon, O. and Chao, L., 'Understanding the evolutionary fate of finite populations: the dynamics of mutational effects', *PLoS Biology*, vol. 5, 2007, e94.

Simeone, A., 'Otx1 and otx2 in the development and evolution of the mammalian brain', *EMBO Journal*, vol. 17, 1998, pp. 6790–98.

Simeone, A., Acampora, D., Gulisano, M., Stornaiuolo, A. and Boncinelli, E., 'Nested expression domains of four homeobox genes in developing rostral brain', *Nature*, vol. 358, 1992, pp. 687–90.

Simeone, A., Acampora, D., Mallamaci, A., Stornaiuoto, A., D'Apice, M. R. and Nigro, V., 'A vertebrate gene related to orthodenticle contains a homeodomain of the bicoid class and demarcates anterior neuroectoderm in the gastrulating mouse embryo', *EMBO Journal*, vol. 12, 1993, pp. 2735–47.

Skinner, B. F., 'Superstition in the pigeon', *Journal of Experimental Psychology*, vol. 38, 1948, pp. 168–72.

Skinner, B. F., *About Behaviorism*, New York, Vintage Books, 1976.

Sober, E., *The Nature of Selection: Evolutionary Theory in Philosophical Focus*, Chicago, IL, University of Chicago Press, 1993.

Sober, E. and Wilson, D., *Unto Others: the Evolution and Psychology of Unselfish Behavior*, Cambridge, MA, Harvard University Press, 1998.

Soschen, A., 'On the nature of syntax', *Biolinguistics*, vol. 2, 2008, pp. 196–224.

Sprecher, S. G. and Reichert, H., 'The urbilateral brain: developmental insights into the evolutionary origin of the brain in insects and vertebrates', *Arthropod Structure and Development*, vol. 32, 2003, pp. 141–56.

Stefani, G. and Slack, F. J., 'Small non-coding RNAs in animal development', *Nature Reviews Molecular Cell Biology*, vol. 9, 2008, pp. 219–30.

Sterelny, Kim, and Griffiths, Paul, E., *Sex and Death: an Introduction to Philosophy of Biology*, Chicago, University of Chicago Press, 1999.

Sterelny, K. and Griffiths, P. E., *Sex and Death: An Introduction to the Philosophy of Biology*, Chicago, IL, University of Chicago Press, 1999.

Strawson, G., *Consciousness and its place in nature*, New York, NY, Oxford University Press, 2006.

Stromswold, K., 'Why aren't identical twins linguistically identical: genetic, prenatal and postnatal factors', *Cognition*, vol. 101, 2006, pp. 333–84.

Stuart-Fox, D. and Moussalli, A., 'Selection for social signalling drives the evolution of chameleon colour change', *PLoS Biology*, vol. 6, 2008, pp. 22–29.

Suda, Y., Kurokawa, D., Takeuchi, M., Kajikawa, E., Kuratani, S., Amemiya, C. *et al.* 'Evolution of Otx paralogue usages in early patterning of the vertebrate head', *Developmental Biology*, vol. 325, 2009, pp. 282–95.

Sultan, S. E. and Bazzaz, F. A., 'Phenotypic plasticity in *Polygonum persicaria* II: norms of reaction to soil moisture and the maintenance of genetic diversity', *Evolution*, vol. 47, 1993, pp. 1032–49.

Suzuki, D. T., Griffiths, A. J. F. and Lewontin, R. C., *An Introduction to Genetic Analysis*, 2nd edition, San Francisco, CA, W. H. Freeman, 1981.

Tariq, M., Nussbaumer, U., Chen, Y., Beisel, C. and Paro, R., 'Trithorax requires hsp90 for maintenance of active chromatin at sites of gene expression', *Proceedings of the National Academy of Sciences USA*, vol. 106, 2009, pp. 1157–62.

Teo'tonio, H., Chelo, I. M., Bradic, M., Rose, M. R. and Long, A. D., 'Experimental evolution reveals natural selection on standing genetic variation', *Nature Genetics*, vol. 41, 2009, pp. 251–57.

Theissen, G., 'Saltational evolution: hopeful monsters are here to stay', *Theory in Biosciences*, vol. 128, 2009, pp. 43–51.

Thom, R., *Structural Stability and Morphogenesis*, Reading, MA, W. A. Benjamin, 1975.

Thompson, D. W., *On Growth and Form*, Cambridge, UK, Cambridge University Press, 1917 (abridged edition). Reprinted and edited by John Tyler Bonner, New York, Dover, 1992.

Todd, P. and Gigerenzer, G., 'Bounding rationality to the world', *Journal of Economic Psychology*, vol. 24, 2003, pp. 143–65.

Tolman, E. C., 'Cognitive maps in rats and men', *Psychological Review*, vol. 55, 1948, pp. 189–208.

Tooby, J. and Cosmides, L., 'Evolutionary psychology: conceptual foundations', in D. E. Buss (ed.), *Handbook of Evolutionary Psychology*, Hoboken, NJ, Wiley, 2005, pp. 5–67.

Tooby, J., Cosmides, L. and Barrett, C. H., 'The second law of thermodynamics is the first law of psychology: evolutionary developmental psychology and the theory of tandem, coordinated inheritances' [comment on Lickliter and Honeycutt], *Psychological Bulletin*, vol. 129, 2003, pp. 858–65.

Trepat, X., Deng, L., An, S. S., Navajas, D., Tschumperlin, D. J., Gerthoffer, W. T. *et al.*, 'Universal physical responses to stretch in the living cell', *Nature*, vol. 447, 2007, pp. 592–95.

Trevisan, M. A., Mindlin, G. B. and Goller, F., 'Nonlinear model predicts diverse respiratory patterns of birdsongs', *Physical Review Letters*, vol. 96, 2006, pp. 1–4.

True, H. L., Berlin, I. and Lindquist, S. L., 'Epigenetic regulation of translation reveals hidden genetic variation to produce complex traits', *Nature*, vol. 431, 2004, pp. 184–89.

Trut, L. N., 'Early canid domestication: the farm-fox experiment', *American Scientist*, vol. 87, 1999, pp. 160–69.

Turing, A. M., 'The chemical basis of morphogenesis', *Philosophical Transactions of the Royal Society B of London. Series B. Biological Sciences*, vol. 237, 1952, pp. 37–72.

Venditti, C. and M. Pagel 'Speciation as an active force in promoting genetic evolution', *Trends in Ecology and Evolution*, vol. 25 (1), 2009, pp. 14–20.

Vercelli, D., 'Genetics, epigenetics and the environment: switching, buffering, releasing', *Journal of Allergy and Clinical Immunology*, vol. 113, 2004, pp. 381–86.

Vercelli, D., 'Discovering susceptibility genes for asthma and allergy', *Nature Reviews in Immunology*, vol. 8, 2008, pp. 169–82.

Volterra, V., *Leçons sur la Théorie Mathématique de la Lutte pour la Vie*, Paris, Gauthier-Villars, 1931.

von Frisch, K., *The Dance Language and Orientation of Bees*, Cambridge, MA, Harvard University Press, 1967.

von Mutius, E., 'Allergies, infections and the hygiene hypothesis – the epidemiological evidence', *Immunobiology*, vol. 212, 2007, pp. 433–39.

Waddington, C. H., 'The genetic basis of the 'assimilated bithorax' stock', *Journal of Genetics*, vol. 55, 1956, pp. 241–45.

Waddington, C. H., *The Strategy of the Genes*, London, Routledge, 1957.

Wagner, A., *Robustness and Evolvability in Living Systems*, Princeton, NJ, Princeton University Press, 2005.

Wang, E. T., Sandberg, R., Luo, S., Khrebtukova, I., Zhang, L., Mayr, C. *et al.*, 'Alternative isoform regulation in human tissue transcriptomes', *Nature*, vol. 456, 2008, pp. 470–76.

Watson, J. and Crick, F., 'The structure of DNA', in *Cold Spring Harbor Symposia on Quantitative Biology*, vol. 18, 1953, pp. 123–31.

Wen, Q. and Chklovskii, D. B., 'Segregation of the brain into grey and white matter: a design minimizing conduction delays', *PLoS Computational Biology*, vol. 1, 2005, 617–30.

Wesson, R., *Beyond Natural Selection*, Cambridge, MA, The MIT Press, 1991.

West, G., Brown, J. and Enquist, B., 'A general model for the allometric scaling laws in biology', *Science*, vol. 276, 1997, pp. 122–26.

West, G. B., Brown, J. H. and Enquist, B. J., 'The fourth dimension of life: fractal geometry and allometric scaling of organisms', *Science*, vol. 284, 1999, pp. 1677–79.

West, G., Woodruff, W. H. and Brown, J., 'Allometric scaling of metabolic rate from molecules and mitochondria to cells and mammals', *Proceedings of the National Academy of Sciences USA*, vol. 99, 2002, pp. 2473–78.

West-Eberhard, M.-J., *Developmental Plasticity and Evolution*, Oxford, UK, Oxford University Press, 2003.

West-Eberhard, M.-J., 'Developmental plasticity and the origin of species differences', *Proceedings of the National Academy of Sciences USA*, vol. 102, 2005, pp. 6543–49.

Williams, F., *Ampulex compressa* (Fabr.), a cockroach-hunting wasp introduced from New Caledonia into Hawaii', *Proceedings of the Hawaiian Entomological Society*, vol. 11, 1942, pp. 221–33.

Wilson, M. D., Barbosa-Morais, N. L., Schmidt, D., Conboy, C. M., Vanes, L., Tybulewicz, V. L. J. *et al.*, 'Species-specific transcription in mice carrying human chromosome 21', *Science*, vol. 322, 2008, pp. 434–38.

Wilson, M., Daly, M. and Weghorts, S. J., 'Household composition and the risk of child abuse and neglect', *Journal of Biosocial Science*, vol. 12, 1980, pp. 333–40.

Wimsatt, W., 'False models as means to truer theories', in M. Nitecki and A. Hoffman (eds), *Neutral Models in Biology*, London, Oxford University Press, 1987, pp. 23–55.

Wimsatt, W. C., 'Generative entrenchment and the developmental systems approach to evolutionary processes', in S. Oyama, P. E. Griffiths and R. D. Gray (eds), *Cycles of Contingency: Developmental Systems and Evolution*, Cambridge, MA, MIT Press, 2003, pp. 219–38.

Woese, C. R. and N. Goldenfeld, 'How the Microbial World Saved Evolution from the Scylla of Molecular Biology and the Charybdis of the Modern Synthesis.' *Microbiology and Molecular Biology Reviews*, vol. 73, 1 March, 2009, pp. 14–21.

Wright, R., *The Moral Animal: Evolutionary Psychology and Everyday Life*, New York, Vintage Books, 1994.

Wright, S., 'Evolution in Mendelian populations', *Genetics*, vol. 16, 1931, pp. 97–159.

Wright, S., 'The roles of mutation, inbreeding, crossbreeding, and selection in evolution', in *Proceedings of the Sixth Congress on Genetics*, Ithaca, New York, 1932, pp. 356–66.

Yang, A. S., 'Modularity, evolvability, and adaptive radiations: a comparison of the hemi- and holometabolous insects', *Evolution & Development*, vol. 3, 2001, pp. 59–72.

Yauk, C., 'Monitoring for induced heritable mutations in natural populations: application of minisatellite DNA screening', *Mutation Research*, vol. 411, 1998, pp. 1–10.

Zelditch, M. L., Wood, A. R., Bonett, R. M. and Swiderski, D. L., 'Modularity of the rodent mandible: integrating bones, muscles, and teeth', *Evolution & Development*, vol. 10, 2008, pp. 756–68.

Zhao, Z., Fu, Y. X., Hewett-Emmett, D. and Boerwinkle, E., 'Investigating single nucleotide polymorphism (SNP) density in the human genome and its implications for molecular evolution', *Gene*, vol. 312, 2003, pp. 207–13.

INDEX

selective breeding model, 99,
115–116, 119, 155–156
self-organization, xviii, 74–75,
91–92
Selfish Gene concept, 120, 156,
163
sex determination in reptiles,
70–71
sexual selection, 176–177
sheep, 62, 155
shifting equilibria, 20–21, 185
signalling pathways, 34–35,
47–48
single-level theories, xxi–xxii,
158, 221
size and fitness, 124, 126–127,
134
Skinner, Burrhus Frederic, xviii,
3–4
see also behaviourism;
learning theory; operant
conditioning theory
Sober's sieve (Elliot Sober),
127–130
spandrels analogy
anticipated, 45, 86
developed, 99–101, 103,
110–111, 114–115
explained, 96–98
speciation
Darwinian view of, 2, 7
gene expression and, 64
rapid changes in, 51–52
species
formal classification, 191

genealogy of, 1, 179
master gene conservation
across, 28–29
as small subset of possible
diversity, 92
see also body plan
evolution; phenotypes
spiders, 91, 140, 149
split response experiments,
104–105
split stimulus experiments,
101–103, 131, 154
starting assumptions, evolution
and learning theory, 6
Stich, Stephen, 173
stimulus-response associations
(S-R)
effective stimulus problem,
101–102, 206
psychological traits as, 4
sugar digestion in E. coli,
60–62
Swiss apple fallacy, 123–127
systemic modules, 48–49

T

teleological explanations, 96–
98, 100, 110, 122, 205–206
temperature
control by insect wings, 87
sex determination and,
71–72
theories
single- and multi-level, xxi–
xxii, 158, 221